12th International Conference on Advanced Manufacturing Technologies (ICAMaT 2022)

IOP Conference Series: Materials Science and Engineering
Volume 1268

Bucharest, Romania
20 – 21 October 2022

ISBN: 978-1-7138-6427-1
ISSN: 1757-8981

Printed from e-media with permission by:

Curran Associates, Inc.
57 Morehouse Lane
Red Hook, NY 12571

proceedings.com

Some format issues inherent in the e-media version may also appear in this print version.

Institute of Physics
Dirac House, Temple Back
Bristol BS1 6BE UK

Phone: 44 1 17 929 7481
Fax: 44 1 17 920 0979

techtracking@iop.org

Additional copies of this publication are available from:

Curran Associates, Inc.
57 Morehouse Lane
Red Hook, NY 12571 USA
Phone: 845-758-0400
Fax: 845-758-2633
Email: curran@proceedings.com
Web: www.proceedings.com

TABLE OF CONTENTS

IOP Conf. Series: Materials Science and Engineering 1268 (2022) 011001 doi:10.1088/1757-899X/1268/1/011001

Preface

The 12th International Conference of Advanced Manufacturing Technologies, ICAMaT 2022, was held in the University POLITEHNICA of Bucharest, Romania, Faculty of Industrial Engineering and Robotics, on 21th of October 2022, with the remarkable participation of experts from Algeria, Spain, Slovenia and Romania.

The first ICAMaT Conference was held in 1999 under the acronym CITAF (derived from the abbreviation in Romanian) at the initiative of the Manufacturing Engineering Department (TCM) and had a national character. The following Conference sessions were held almost every two years and attracted increasingly more specialists.

Since 2005, with the clear prospect of Romania's accession to the European Union, the conference takes on an international character and changes its name to ICAMaT - International Conference on Advanced Manufacturing Technologies.

After an online edition (ICAMaT 2020) and a year (2021) when the conference could not be held due to the challenges posed by the COVID pandemic and other events unfolding in Europe, ICAMaT returned to the scientific arena with a set of 20 papers that were presented by researchers from universities, research institutes as well as private enterprises, such as: University POLITEHNICA of Bucharest; "Gheorghe Asachi" Technical University of Iaşi; Technical University of Cluj-Napoca; Iuliu Hatieganu University of Medicine and Pharmacy of Cluj-Napoca; "Transilvania" University of Brașov; University Kasdi Merbah, Ouargla, Algeria; University of Maribor, Slovenia; GHI Hornos Industriales S.L., Biscay, Spain; INCAS – National Institute for Aerospace Research "Elie Carafoli", Bucharest; National Institute of Research and Development in Mechatronics and Measurement Technique Bucharest (INCDMTM); Nuclear NDT Research and Services, Bucharest and Dr. Köcher S.R.L., Bucharest.

26 papers were submitted for the conference, 24 being selected for review and 20 being accepted and presented. The Editors are thankful to the authors for their valuable contributions and to the reviewers for their timely reviews and recommendations. We also wish to thank the members of the Organizing Committee for the consistent and substantial effort they undertook in organizing this conference.

The conference papers were split in 3 sections, 2 of the sections were held in person and the last section was held online due to several requests made by authors from research institutions outside of Romania.

The organizers used the Microsoft Teams platform belonging to the University POLITEHNICA of Bucharest, where a Teams group, named *ICAMaT 2022 - S3. Industrial Products & Systems - online section*, was created. A link to the group was sent to all participants and it was also displayed on the conference program which was uploaded on the conference website. Invitations for joining the team, using the Microsoft Teams specific tools, were sent to all participants several days before the conference's start.

The time recommended for each paper presentation was a maximum of 25 minutes for the papers presented in person and a maximum of 20 minutes for the papers presented online, but it was up to the sections' chairs to allow for time extensions, if requested. Questions and comments followed the presentation of each paper. The presentations and discussions of the online section were recorded and are available on the Teams platform for all participants.

There were no major technical problems during the conference, the participants were able to attend and understand the presentations. Both the authors and the other attendees succeeded in acquiring useful and relevant scientific information from their participation.

Manuela-Roxana DIJMĂRESCU, Tom SAVU, Ioan-Cristian TARBĂ - Editors

Scientific Committee

IOP Conf. Series: Materials Science and Engineering 1268 (2022) 011001 doi:10.1088/1757-899X/1268/1/011001

Assoc.prof. **Lucian LĂZĂRESCU**, Technical Univ. of Cluj-Napoca, Romania
Prof. **Marion MARTINY**, Université de Lorraine - Metz, France
Prof. **Traian MAZILU**, Univ. POLITEHNICA of Bucharest, Romania
Prof. **Sebastien MERCIER**, Université de Lorraine - Metz, France
Prof. **Eduard Laurențiu NIȚU**, Univ. of Pitesti, Romania
Prof. **Gheorghe OANCEA**, Univ. "Transilvania" of Brasov, Romania
Assoc.prof. **Răzvan-Ioan PĂCURAR**, Technical Univ. of Cluj-Napoca, Romania
Assoc.prof. **Gabriela Adriana PLĂIAȘU**, Univ. of Pitesti, Romani
Prof. **Hanan SALEET**, Applied Science Private University, Amman, Jordan
Prof. **Tom SAVU**, Univ. POLITEHNICA of Bucharest, Romania
Assoc.prof. **Jeremie SCHUTZ**, Université de Lorraine - Metz, France
Prof. **Elena SCUTELNICU**, Univ. "Dunarea de Jos" of Galati, Romania
Prof. **Hongwu SONG**, Institute of Metal Research, Chinese Academy of Science, Shenyang, China
Assoc.prof. **Paulina SPÂNU**, Univ. POLITEHNICA of Bucharest, Romania
Prof. **Nicolae-Doru STĂNESCU**, Univ. of Pitesti, Romania
Lect. **Ioan - Cristian TARBĂ**, Univ. POLITEHNICA of Bucharest, Romania
Lect. **Petre Gheorghe TIRIPLICĂ**, Univ. POLITEHNICA of Bucharest, Romania

Organizing Committee

Lect. **Manuela-Roxana DIJMĂRESCU**, Univ. POLITEHNICA of Bucharest, Romania
Prof. **Andrei DUMITRESCU**, Univ. POLITEHNICA of Bucharest, Romania
Assoc.prof. **Ionuț GHIONEA**, Univ. POLITEHNICA of Bucharest, Romania
Ing. **Alexandru Bogdan JUGRAVU**, Univ. POLITEHNICA of Bucharest, Romania
Prof. **Tom SAVU**, Univ. POLITEHNICA of Bucharest, Romania
Lect. **Ioan-Cristian TARBĂ**, Univ. POLITEHNICA of Bucharest, Romania

Peer Review Statement

All papers published in this volume have been reviewed through processes administered by the Editors. Reviews were conducted by expert referees to the professional and scientific standards expected of a proceedings journal published by IOP Publishing Publishing.

- **Type of peer review:** Double Anonymous
- **Conference submission management system:** Morressier
- **Number of submissions received:** 26
- **Number of submissions sent for review:** 24
- **Number of submissions accepted:** 20
- **Acceptance Rate (Submissions Accepted / Submissions Received × 100):** 76.9
- **Average number of reviews per paper:** 2
- **Total number of reviewers involved:** 20
- **Contact person for queries:**
  Name: Manuela-Roxana Dijmarescu
  Email: manuela.dijmarescu@upb.ro
  Affiliation: Polytechnic University of Bucharest - Manufacturing Engineering

IOP Conf. Series: Materials Science and Engineering 1268 (2022) 012001 doi:10.1088/1757-899X/1268/1/012001

Digital transformation of the laboratories

Miruna Lucretia Comeaga[1]

[1]University POLITEHNICA of Bucharest, Faculty of Mechatronics and Mechanical Engineering, Department of Mechatronics and Precision Mechanics, 313 Spl. Independentei st., 060042, Bucharest, Romania

miruna.comeaga@gmail.com

Abstract. The scientific world is evolving; laboratory processes that were once entirely manual are being optimized through automation and data connectivity. While documents previously had to be printed, signed, and physically stored in boxes, the recent developments of data management systems and software platforms allow scientists to sign and store electronic records in a secure and future-proof format. Advanced analytics solutions, such as artificial intelligence and machine learning, are being used to connect, collate and analyse data from different facilities around the globe. Digital transformation is occurring in many aspects of the everyday life, from the ability to control household appliances through apps, to monitoring our health through wearable devices. But it is not just happening at home: the fourth industrial revolution is already in full expansion and digital technologies connect automated processes and equipment, monitor and control supply chains and work alongside robots programmed to leverage artificial intelligence (AI). The decision to digitally transform the laboratory is not without its challenges. New methodologies need to be evaluated and approved by the relevant regulatory body. The diverse scope of applications around the world makes it difficult for organizations looking to adopt such capabilities, despite the potential benefits. In addition to exploring the benefits and challenges of digital transformation, this article will show how new and practical solutions can be integrated to increase laboratory productivity and accelerate scientific research.

1. Introduction

The scientific world is evolving; laboratory processes that were once entirely manual are being optimized through automation and data connectivity. While documents previously had to be printed, signed and physically stored in boxes, the recent developments of data management systems and software platforms allow scientists to sign and store electronic records in a secure and future-proof format. Advanced analytics solutions, such as artificial intelligence and machine learning, are being used to connect, collate and analyze data from different facilities around the globe.

Digital transformation is occurring in many aspects of everyday life, from the ability to control household appliances through apps, to monitoring our health through wearable devices. But it is not just happening at home: the fourth industrial revolution is already in full expansion and digital technologies connect automated processes and equipment, monitor and control supply chains and work alongside robots programmed to leverage artificial intelligence (AI) [1]. This transformation, known as digitalization, differs from digitization, which is the process of converting information into a digital format. Digitization refers to the transition from analog to digital data, for example, replacing paper records with digital records, using electronic tracking systems such as laboratory notebooks (ELN), and creating an electronic database of standard operating procedures (SOPs). Simply put, digitization refers to information, while digitalization refers to the physical processes that enable conversion to digital technologies; for example, connecting instruments, reconciling software and data, harmonizing processes and workflows and integrating other laboratory components to achieve a complete digital transformation. Digitalization means using digital technology to collect data, set trends, automate processes and make better business decisions [2].

The decision to digitally transform the laboratory is not without its challenges. New methodologies need to be evaluated and approved by the relevant regulatory body. The diverse scope of applications around the world makes it difficult for organizations looking to adopt such capabilities, despite the potential benefits. This article is presenting some leading digital technologies that can be integrated to increase laboratory productivity and accelerate scientific research. In addition, it also explores the considerations that need to be considered, showcasing the benefits and challenges of the digital transformation strategy.

2. Digital technologies integrated in laboratories

Digital transformation can be widely defined as the process of integrating digital technology to create or change business processes, operations, culture, and improve organizational results, deliverables and customer experiences. Digital transformation involves a profound change in the management of technology, data and associated processes. Increased use of digital technologies creates unique opportunities; tools such as the internet of things (IoT) and machine learning (ML) are radically changing processes and workflows in labs. In addition, the adoption of these technologies is essential to ensure that all tools, operations and workers are connected [3].

Laboratories must transform the way they create, manage and store data to ensure a seamless flow of information. Because increased connectivity can create challenges across multiple regulatory regimes, laboratories must develop comprehensive strategies that comply with regulatory guidelines to enforce and maintain data security. Increased use and reliance on electronic records, laboratory information systems and digital applications is generating significantly large volumes of data and strengthening the role of artificial intelligence (AI) across the scientific field. For example, in terms of laboratory efficiency and sustainability, AI can be used to identify areas of waste, improve processes and protocols and enable more rational ordering of laboratory tests [4].

Given their impact on organizations, digital enablers should be strong enough to support the full lifecycle of highly regulated processes while providing flexibility to adapt to varying workflow and data management needs. Some leading digital technologies are discussed below [5] [6] [7]:

- IoLT: refers to devices used to collect and share data over the internet. IoLT (Internet of Laboratory Things) can be used in the continuous instruments and parameters monitoring, inventory management and data collection. Reducing manual tasks and consequently lowering the risk of human errors increases productivity and reduces costs.
- Cloud computing: refers to remote technologies (e.g., servers, databases, software, analytics) that store, manage and centralize data and connect disparate tools in different locations and organizations.
- Informatics: refers to computational systems that process data for storage and retrieval. For example, laboratory information management system (LIMS) and electronic lab notebooks (ELN) can be used together to track samples from receipt to reporting of result and secure the introduction of methodological data as well as metadata to the digital ecosystem. Usually, LIMS is best suited to handle more structured data, such as in a manufacturing or process environment, while ELN is generally applicable to less structured data, often in a discovery or research application.
- Scientific data management systems (SDMS): provide a centralized database for collecting, organizing and storing data. Data sources can be diverse (from any analytical tools or equipment) and are restructured to facilitate data analysis. With its ability to generate audit logs and diligent reports, SDMS is critical for quality control, ensuring data integrity and thus

compliance with regulatory guidelines. A state-of-the-art SDMS can provide connectivity by connecting to instrument sensors and eliminating manual data entry via devices such as electronic lab notebooks or advanced tools such as augmented reality but also compliance and security, through a secure system with data access control, version control and automated logging of events, ensuring all raw data and associated metadata is accessible for the entire data lifecycle.

- Artificial intelligence (AI): refers to "smart" technologies which perform tasks that usually require human intelligence (e.g., decision-making). AI not only enables the live processes monitoring and data interpretation, but also real-time troubleshooting of all the networked systems. With predictive and reactive intelligence, AI can be used as a holistic lab manager for many tasks including inventory management, equipment maintenance and supply chain activities. Regarding the laboratory efficiency and sustainability, AI can be used to identify areas of waste, improve processes and protocols and enable more rational laboratory test ordering.
- Automation: the switch to integrate modular automation systems not only improves the reliability and reproducibility of processes, workflows and data and sample management but enables these tasks to be performed with minimal intervention from laboratory staff.

3. Important considerations for a digital transformation strategy

The move to paperless systems coupled with the pressure to integrate data-driven technologies is forcing laboratories to become 'smarter'. There are many factors that influence how the laboratory environment will change over time that should be considered as part of the digital transformation strategy [8] [9].

- People and culture

The future lab will place increasing emphasis on global and inter-organizational cooperation. Researchers will become more familiar with the work of others, removing barriers to collaboration and increasing the adoption of new methods as they become available. Importantly, a significant proportion of manual work will be eliminated with the adoption of robotics and automation, allowing scientists to focus on the key aspects and complex facets of their work.

- Process evolution and development

The integration of robotics and automation will continue to increase the efficiency and overall quality of laboratory processes and experiments, by replacing a significant proportion of manual tasks. The digital lab will also be increasingly driven by environmental and sustainability objectives; experiments that currently involve toxic and harmful materials will be redesigned to minimize their usa during both the experimental and clean-up phases.

- Data security and management

The availability of accessible, affordable and connected technology allows laboratories to optimize their operations and combine instruments and data more efficiently. This inevitably generates large amounts of data from different sources as close to real-time as possible.

Data needs to be harmonized and standardized before it can be optimally utilized; the use of FAIR (findable, accessible, interoperable, reusable) data principles facilitates the navigation, management, analysis and use of internal and external data.

- Lab environment

The future lab is under pressure to design more open working areas that encourage collaboration and teamwork, rather than the traditional culture of small groups and departments. In addition, the labs of the future can minimize their impact on the environment while increasing efficiency and reducing costs by focusing on sustainability. This includes using energy-efficient instruments and equipment, sourcing more environmentally responsible products and considering eco-friendly methods of manufacturing, packaging and shipping.

- Emerging solutions

Future labs will benefit from the development of innovative technologies. For example, in addition to automation, integrating speech recognition, augmented reality and biometric authentication into experimental workflows will increase the consistency of processes and improve data quality and security by reducing human interaction and automating data capture.

Many labs combine advanced analytics with automation and robotics for faster and more targeted scientific discovery. For example, labs can use robotics and automated systems to generate high-quality data, which is interpreted by machine learning (ML) and used by artificial intelligence (AI) to select the next optimal experiment to perform.

4. Key steps of a digital transformation strategy

There are three main steps to driving digital transformation, as following [10]:

1.The first involves connecting everything to make data available. By taking a step back to look at the existing infrastructure, labs will be able to determine where the problems lie. Having high-quality FAIR data that can be reused with confidence is also a key point. Connecting both experimental and operational data will increase data integrity, compliance and productivity. Connected laboratories are the fundamental layer of the digital transformation.

2.The second step involves automating end-to-end, application-centric workflows to maximize efficiency, increase reproducibility and finally enable labs to adapt to changing workflows and technology so that scientists can focus on the tasks that bring value. Automated laboratories are the second layer of the digital transformation.

3.The last step of the digital transformation journey is the transition to a smart laboratory; one that uses data and advanced analytics tools (e.g. ML and AI), as well as immersive technologies (e.g. hands-free voice accessibility and virtual reality systems) to make informed decisions about data and downstream activities to continuously improve processes and accelerate science. These laboratories represent the third and final layer of the digital transformation.

5. Benefits of digital transformation

While a certain amount of disruption is required to transform an analog lab into a fully digital, connected and "smart" one, the organizational benefits significantly exceed the challenges [5].

- Increased efficiency and productivity - digital transformation enables scientists to communicate within groups and across departments, share real-time data continuously across the lab and stay in touch with evolving customer requirements. In addition, augmented reality, speech recognition technology and artificial intelligence can be used to access software and systems without having to interrupt lab work to record actions or check the next steps. This further enhances data and process integrity, improving efficiency while saving time, money and resources.
- Increasing transparency - automating data workflows increase transparency by identifying trends in processes and practices that lead to small, gradual and consistent improvements in all aspects of the lab.
- Cost savings - immersive solutions enable more efficient collaboration both within and between labs. This makes training, troubleshooting and maintenance much easier for staff, in the end saving costs by improving runtime of tools and equipment.
- Improved profitability - minimizing costs, identifying inefficiencies and improving best practice will lead to increased revenue - 80% of labs report that digital transformation has led to improved profitability.
- Improved customer experience - technological innovations enable organizations to collect and analyze customer data to improve existing products and create new products and services.
- Increased flexibility and resilience - the COVID-19 pandemic has placed increased emphasis on the need for labs to be versatile and resilient. For many, this has involved the adoption of data analytics, automation and other digital tools that have enabled them to use and share online resources on demand to keep their businesses moving forward.
- Better inventory management - an integrated inventory management process provides visibility by connecting valuable information about raw materials, suppliers, delivery methods, and

IOP Conf. Series: Materials Science and Engineering 1268 (2022) 012001 doi:10.1088/1757-899X/1268/1/012001

ultimately the customer experience. Integrated systems that monitor resource usage and provide the ability to reorder stock as needed enable labs to manage advanced acquisitions.

6. Current approach vs. possible approach

In the following table, a list of current approaches based on the standard requirements from ISO/IEC 17025:2017 are presented.

Starting from them and taking into account the digital technologies already introduced, possible approaches for each requirement are highlighted.

No.	ISO/IEC 17025:2017 standard requirements [11]	Current approach	Comments regarding possible approaches based on the standard requirements
1.	The laboratory is responsible for carrying out its testing and calibration activities so that: - it meets the international standard requirements; - it meets the needs of the customers; - it meets the needs of the regulators or the organizations granting recognition.	Conduct periodic audits based on documentations analysis communicated by the laboratory and on-site analysis. The activity of the laboratory cannot be monitored by the client/regulatory body except through records, inherently limited (e.g., memory of analytical equipment, test reports).	The constant acquisition of data relative to the flow of the samples, the operations carried out on them, the equipment and the conditions used allow an automatic analysis of the quality of the process, including by automated systems (such as AI) that automatically generate real-time reports and information to the client/authority.
2.	The management system must cover the activities carried out in the laboratory permanent premises, outside the laboratory permanent premises, in premises used temporary or within mobile laboratories.	The analysis activity tends to be decentralized and it is currently difficult to track all locations by computer flow.	In this case, a possible approach would be if tracking is done automatically through specialized reports directly to the management system, dedicated per topic (HR, testing, validation, etc.). The problem is greatly complicated by the trend towards ultra-distributed systems consisting of micro-test sites or even local sample collection and processing systems (fixed or mobile bins in which a wide range of analyses are performed).
3.	If the laboratory is part of an organization that carries out activities other than testing and/or calibration, the responsibilities of the key personnel who are involved or has influence on the testing and/or calibration activities of the laboratory shall be defined in advance to identify potential conflicts of interest. If the	Statements and training from management guaranteeing that the parent organization will not exert pressure of any kind on the laboratory or staff and/or will allow them full freedom in technical and scientific matters. All statements are accurate. The deviations	Interaction only through computer system, with limited rights and standard messages. Use of specialized software to analyze information and identify attempts at influence. Large-scale use of automated measurement systems that accumulate data in clouds under the supervision of regulatory bodies and without any possibility of fraudulent deletion (changes to accidental errors are possible but

	laboratory wishes to be recognized as a third-party laboratory, it should be able to demonstrate its impartiality and that its personnel is not subject to any commercial, financial or other pressures that could influence the technical decision of the laboratory.	are difficult to identify and are mostly done verbally.	with limited rights and with logging of all changes). Use of long-term data analysis software that can highlight negative trends and the emergence of strange "corrections" of values in a positive direction.
4.	The laboratory must have a management and a deputy for the management and also technical staff with the authority and resources to carry out the tasks and identify the deviations from the quality system or procedures and to initiate actions for preventing or minimizing such deviations.	Staff training and decisions to establish replacements (often without extensive experience due to the current staffing crisis and the difficulty of identifying highly experienced employees).	The use of management "clones" in decision making process, capable of supporting current management activity but also acting as deputies for current tasks.
5.	The laboratory must have policies and procedures in place to ensure the protection of their confidential information and proprietary rights of its clients, including procedures for the protection of electronic storage and transmission of the results.	Rules and procedures backed up by contracts with specialized IT companies but usually without experience in anti-spying/anti-hacking activities because they are very expensive.	Use of specialized cyber-ware protection software based on advanced AI techniques and ways to manage data in the cloud.
6.	The laboratory must specify the responsibility, authority and mutual relationships for all personnel which conduct, perform or verify activities that influence the quality of testing and/or calibration.	Very extensive statements/decisions and job descriptions as staff is limited and must fulfil numerous technical and administrative tasks.	Eliminate time-consuming requirements (e.g., sampling and sample preparation, complicated reporting, statistical analysis of data trends to ensure reliability). Staff become a supervisor/configurator of these automated systems. Authority is governed through the structure of the information system. Individual relationships (e.g., between sampler and analysis or post-processing staff) are eliminated.
7.	The laboratory shall ensure adequate supervision of personnel carrying out testing and calibration activities including personnel undergoing	Periodic reports by the laboratory manager. They are often formal, relationships are already informal or even very close, evaluations are	Supervision is automatic and comprehensive, professional results can be automatically evaluated and as often as needed through online tests including in the virtual environment, errors are quickly

	training by those experienced in methods and procedures used by the laboratory.	exclusively positive. Evaluation is rarely done on simple tests due to staff and time shortage.	identified through trend analysis. Interactions and personal influences are limited between assessor and assessed.
8.	The laboratory must have a technical management that has full responsibility for the technical activities and for providing the resources necessary to ensure the quality required for the laboratory's activities. Technical management refers to the allocation of staff resources, knowledge, skills, methods and procedures for maintaining and improving the standard of analytical services provided by the laboratory.	Appointing a person with a very good CV as laboratory manager and transferring the responsibility to him/her by management issued decision. The laboratory activity is blocked when the person is temporarily unavailable or leaves, he/she is often overloaded with technical administrative activities and "traumatized" by the very high legal responsibility.	The technical manager is much flexible in relation to the current activity that is managed by automated systems and focuses on identifying opportunities for improvement. He/she configures the functions of the technical management AI, validates it together or supervised by a specialized auditor and randomly checks the results.
9.	The laboratory must designate as Quality Manager (QM), a member of the laboratory staff who, irrespective of other duties and responsibilities, has the defined responsibility and authority to ensure that the quality system is always implemented and adhered to. The responsibilities of the QM include conducting internal audits, maintaining adequacy of the quality system, updating documents, identifying quality issues and initiating corrective actions.	An appointment decision to a person with background in quality assurance. Basically, he/she is overloaded with paperwork, authority is relative due to peer relationships. Internal audits are done as infrequently as possible and often formally. Corrective actions are minor and sometimes cannot be taken due to lack of resources (especially planning).	The QM becomes a beneficiary of the reports generated by automated quality analysis systems and a "coach" of them. The documents analysis (form and substance) can be done automatically. The internal audit can be generated at any time because it has continuous access to the information and has the means to analyze it automatically. Automated systems can generate preventive reports (e.g., stocks of chemicals, decrease in measuring accuracy of equipment).
10.	The laboratory must establish and maintain procedures for the control of all documents (generated internally or from external sources) that are part of its management system.	Documents are produced and distributed on paper, using a mailing list. Situations are encountered where the distribution list is not automatically updated when staff change, or documents are not updated when the referential changes.	Automated system for sending documents, staff assessment for their understanding, automatic updating together with references. A new activity is added, to analyze the referential and to re-train the AIs according to external changes (method referential or functional structure of the institution).

IOP Conf. Series: Materials Science and Engineering 1268 (2022) 012001 doi:10.1088/1757-899X/1268/1/012001

11.	The laboratory must establish and maintain procedures for the analysis of requests, tenders and contracts. Analysis records must be maintained, including any significant changes.	Often the requests analysis is verbal or through brief documents, sometimes contracts are filled in at the measurement site (environmental/labor protection). Staff interacting with clients has no technical skills and often clients do not know what they want, only where they want to use the results.	Sophisticated IT systems to assist the client in choosing the correct request, including VR and AR presentation of measurement situations and the use of a particular type of result, special software for contextual text analysis of client requirements. Automatic storage of requirements and requests for predictive analysis of next requests.
12.	The laboratory must have a policy and procedure(s) for the selection and procurement of services and supplies used that affect the quality of testing and/or calibrations.	Often the evaluation is rarely done, information from suppliers is biased and commercial, history is difficult to access because laboratory staff do not keep records on the evolution of their reagents' quality.	Evaluation is done automatically based on the data pool during automated measurements, trends in suppliers' quality are identified, social media can be automatically scanned to evaluate public information about suppliers (rating, etc).
13.	The laboratory must be willing to cooperate with clients or their representatives, in clarifying the client's requirements and in monitoring the laboratory's performance in relation to the work performed, ensuring confidentiality in relation to other clients. Ensuring reasonable access for the customer or its representative to relevant areas of the laboratory to witness tests and/or calibrations carried out for the customer.	It is stated in the procedures but in most cases the cooperation is limited to sending evidence for analysis and then receiving reports. Laboratories do not allow access because the premises are restricted, can be contaminated and there are confidentiality problems between clients.	Controlled access to the sample analysis process (condition, equipment, accuracy) provided by a unique code received at the time of sample reception (taken by the customer or automatically by the laboratory). A warning is automatically generated in case of identification of an anomaly in the sample analysis (e.g., contaminated or unlikely given the previous history of the location) and the sampling is repeated if requested by the customer.
14.	Laboratory management must ensure the competence of all those who work with specific equipment, perform testing and/or calibration, assess the results and sign test reports and calibration certificates. Where trainee personnel are used, appropriate supervision must be	A formal process for recruitment often occurs because staff is selected in a tight labor market, is quickly trained (a few days) for a wide range of methods (leading to a temporary retention of information) and is evaluated once a year.	Most of the repetitive tasks (sampling, measurement, measurement data processing, post-processing and trend analysis) are transferred to automated and robotic systems. Employees are selected through complex analysis processes, including AI. Their training is done gradually in modules and relies heavily on AR/VR, including the

	ensured. Personnel performing specific tasks must be qualified based on relevant education, training, experience and demonstrated ability, as necessary.		operation of virtual equipment in a virtual environment. Assessment is done indirectly, through automatic analysis of employee behavior, and leads to personalized adjustment of subsequent training. Many new jobs are emerging.
15.	The laboratory must be equipped with all is necessary for sampling, measuring and testing equipment required for the correct performance of tests and/or calibrations (including sampling, preparation of test objects and/or calibrations, processing and analysis test and/or calibration data).	Current situations: equipment is calibrated with CRM (certified reference materials) but with reduced frequency or only annually for those calibrated at NIM or similar. No detailed long-term analysis of equipment performance is conducted (it is laborious and requires very competent staff). Thus, the effect of some measurement errors can propagate backwards over a long period of time - reports being cancelled or recalled.	Automatic analysis of the measurement data and correlation with previous sampling history from that location and other data measured on the same sample can reveal sample contamination and equipment malfunctions. Determination of time between calibrations is based on international criteria but with values subjectively chosen by the laboratory. An automatic evaluation of the equipment stress and a prediction of the time between two calibrations can be performed, including predictive maintenance techniques.

7. Conclusions and future trends

In the age of digitalization, scientific labs are also busily gearing up for the future - robotic systems and automation, full digitalization, a flexible, modular work environment, smart materials and functional surfaces - everything is possible.

In recent decades, laboratories have undergone a continuous change, moving from paper-based documentation systems to paperless systems. Not only that, laboratories are also experiencing a steady evolution from electronic systems to data-driven, integrated technologies. In short, laboratories are becoming smarter.

First and foremost, digitizing laboratory data greatly simplifies record keeping. Researchers can easily sort, label, store, access and search digital data. Secondly, data security is one of the greatest benefits offered by digitization. Nowadays, digital data can be encrypted, securely stored in clouds and protected by access control. Moreover, digital data makes it much easier to create a full audit trail for intellectual property protection or regulatory compliance. Digital data platforms can automatically track data entries or edits with time stamps so it is very clear who contributed to new ideas and when intellectual property was generated.

Overall, lab digitalization improves processes, simplifies workflows and makes research and development more efficient. And with big advances in machine learning, deep learning and artificial intelligence, laboratories are becoming smart.

Smart digital laboratories are characterized by an increasing variety of software packages and management systems, which can all be linked to further technologies, such as mobile phones, clouds and the Internet of Things (IoT). Furthermore, artificial intelligence and machine learning facilitate the work with large datasets and supply researchers with an accuracy that has never existed before.

With more and more integrative technologies available today, the key, in the future, will lie in identifying those technologies that are best suited for specific laboratories and understanding how certain aspects of the lab could be improved. By linking all smart technologies in a digital laboratory, data could be fed directly into a smart documentation system, which would greatly lower the risk of human error and streamline workflows. Researchers will then be able to access their data from anywhere at any given point in time, as well as share their data with colleagues all over the world.
The lab of the future is smart, connected and touchless:

- Built-in research workflows: sync all lab notes, reminders, updates, supply lists and more; instantly share documents with lab members, improving collaboration;
- Active data access: step-by-step guide through experimental processes via voice or visual prompts; voice-activated access to recorded experiments, existing data, calendar, to-do lists; add notes directly to each step of the process for faster revisions;
- Handsfree data capture: take notes, record results, create supply lists, set reminders; record notes from anywhere in the lab, streamlining workflow and reducing error-causing distractions.

Implementing a comprehensive digital strategy is key to building a connected scientific ecosystem. It can often require adjustments to company culture and must consider the people, tools, processes and data within an organization. However, in addition to increasing productivity, improving organizational performance and communication, the decision to digitally transform the laboratory will enable the optimization of processes and workflows, increase efficiency and simplify quality control and compliance.

References

[1] https://pages.labvantage.com/rs/856-KCG-620/images/LabVantage-How-AI-Powered-Analytics-Transforms-LIMS-Data-White-Paper.pdf
[2] https://www.oecd.org/gov/regulatory-policy/Regulatory-effectiveness-in-the-era-of-digitalisation.pdf
[3] https://assets.thermofisher.com/TFS-Assets/CMD/Reference-Materials/wp-95975-lab-future-wp95975-en.pdf
[4] http://www.pharmatimes.com/thought_leadership/building_the_digital_lab_of_the_future_1341416
[5] https://keyvatech.com/wp-content/uploads/2020/01/IDC-2020-Futures.pdf
[6] Sisodia A and Jindal R 2021 A meta-analysis of industry 4.0 design principles applied in the health sector *Engineering applications of artificial intelligence* (vol 104) ed Pergamon-Elsevier Science Ltd
[7] Eisen K, Eifert T, Herwig C and Maiwald M 2020 Current and future requirements to industrial analytical infrastructure-part 1: process analytical laboratories *Analytical and bioanalytical chemistry* (vol 412) pp 2027-2035
[8] Imran F, Shahzad K, Butt A and Kantola J 2021 Digital transformation of industrial organizations: toward an integrated framework *Journal of change management* (vol 21) pp 451-479
[9] Lippi G and da Rin G 2019 Advantages and limitations of total laboratory automation: a personal overview *Clinical chemistry and laboratory medicine* (vol 57) pp 802-811
[10] https://resources.experfy.com/future-of-work/four-things-to-consider-when-adopting-lab-automation/
[11] ISO/IEC 17025:2017

Development of an additive fabricated biped mechanical system

Ileana Dugăeşescu [1,a], Gabriela-Marina Pârvu[1,b], Vlad-Cristian Enache[1]

University POLITEHNICA of Bucharest, Faculty of Industrial Engineering and Robotics, Department of Manufacturing, Romania

a) ileana_d1@yahoo.com ; b) gabriela_marina.ene@upb.ro

Abstract. The use of mechatronic components and the incorporation of additive manufacturing into the process of product development have grown in popularity over the past several years. The goal of the research conducted in this paper is to develop a prototype of a monomobile biped mechanical system and to optimize the printing parameters for the additive manufacturing stage. This paper presents the modelling of the components of a biped mechanical system and the simulation of their operation. The modelling step is followed by setting the printing parameters in the dedicated Z-Suite software and creating the G-code for the Zortrax M300 Plus additive manufacturing equipment. Lastly, the paper presents the connection of the sensors to the Plusivo board, the mounting procedure to the 3D printed parts that make up the biped mechanical system as well as the operation methodology of the sensors. The resulting prototype is then used to monitor parameters specific to the sensors used.

1. Introduction

The process of product development has changed significantly in recent years, along with the rate at which additive manufacturing technologies (AM), which have themselves been around for over 30 years, are being adopted. Rapid prototyping is nowadays a predominant part of the product development process because of the benefits of additive manufacturing technology. The newest 3D printing technology is portable and produces actual products quickly.

The monomobile biped mechanical system fills a demand that field-identified customers have. Being a prototype, it assists in tracking some characteristics from the environment (temperature, mobility and gas) for a typical household and to give feedback if thresholds are exceeded.

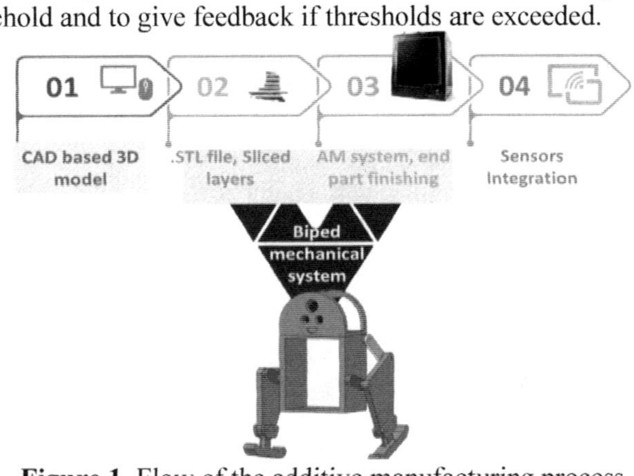

Figure 1. Flow of the additive manufacturing process

For the consumer goods industry-the sector of industry our prototype is dedicated, 3D printing offers significant value. From the ability to create customized products to faster product development cycles, consumer goods companies and retailers can use 3D printing to remain competitive within an ever-changing, consumer-driven market landscape [1,6,7].
Particular attention is placed on integration between additive manufacturing techniques and sensors. The first one enables material swapping throughout the manufacturing process and, as a result, can realize complex systems [5,7]. By integrating predictions from a physical model-the prototype (using simulation on computer software) with sensor signatures, the objective is to identify potentially harmful fluctuations in the domestic environment when various electro-mechanical systems operate.
Additive manufacturing has gone through somewhat of a revolution over the last five to ten years. At one time thought to be a consumer-oriented technology, AM has now evolved into a viable industrial manufacturing solution. One of the most important aspects of mechanical systems constructed using additive manufacturing (AM) is finding the optimal location for mounting the sensors [2, 3,5] such that they can be integrated.

2. System modelling

In order to address the identified needs, the goal of this research is to create a robot prototype utilizing additive manufacturing and integrating sensors. Design for 3D printing in the context of AM, which is also commonly referred to as DfAM or DFAM, represents an important step for the process of manufacturability. The design step makes use of techniques and tools with the purpose of optimizing functional performance, ease of manufacturing and cost (both in terms of time and materials) further showing the relevance of AM.
The kinematic elements of the mechanical biped system were designed using specialized software. Afterwards, the designed parts were assembled and thus obtaining the final mechanical system that is shown in figure 2a. The system front part was then resized and the front part of the body was designed, as to accommodate the LCD display and the buzzer.

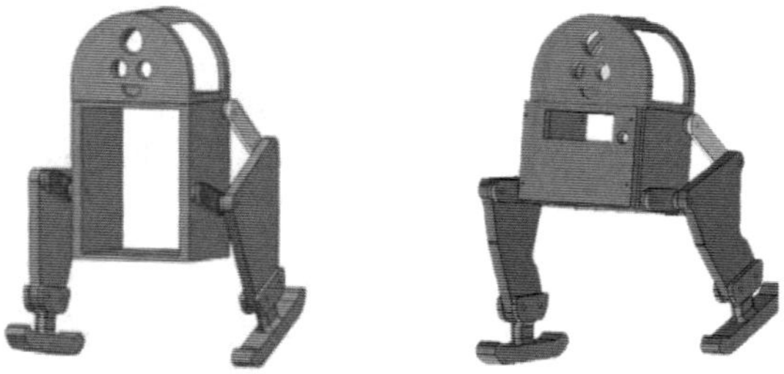

a) The initial design idea for the mechanical system **b)** The final assembly for the mechanical system

Figure 2. The biped mechanical system

A topology optimization study was conducted in order to reduce the weight of the biped mechanical system. The study was created for three different weight reduction percentages: 15%, 30% and 50%. After the optimal reduction percentage was chosen for the 3D printing part, the rest of the biped mechanical system was designed – the rest of the components.

Figure 3. a) 15% mass reduction b) 30% mass reduction c) 50% mass reduction

The upper part of the mechanical biped system was designed so that electrical components can be introduced into it while maintaining a reduced weight of the assembly and due to the principle of the quantity of material used in AM.

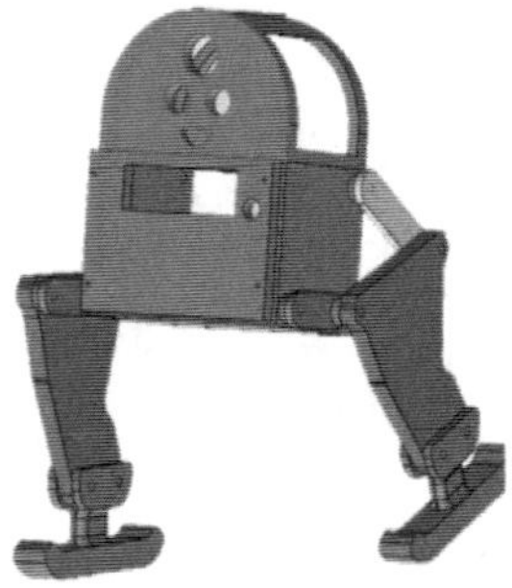

Figure 4. The final model of the biped mechanical system

3. Additive manufacturing and optimization of system printing parameters

The process of conceptualizing, designing, producing, and maintaining items is improving due to additive manufacturing technology (AM), also referred to as 3D printing. AM opens digital design tools, facilitates on-demand, production without specialized tools or equipment and enables exceptional flexibility and breakthrough performance across industries. While some issues, such acquiring enhanced material are characteristic for the early-stage design of the mechanical system, have been on the topic since the advent of additive manufacturing, new approaches have developed in creating the prototype. These topics involve basic science, including materials (see the comparison in Table 1), lightweight structures, bioprinting and integrated electronics.

Figure 5 presents the components of the biped mechanical system prototype that are suitable for additive manufacturing – the rest of the components will be integrated afterwards.

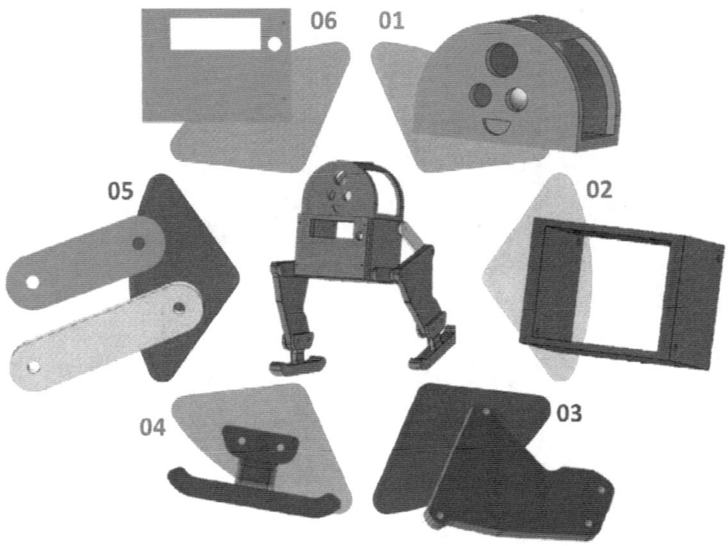

Figure 5. Components of the biped mechanical system additive manufactured

Final material choice for biped the mechanical system is of key importance and is directly correlated with the functionality of the final product, as well as with customer compliance [1]. Based on these factors, the research takes into consideration the performance of three materials manufactured using AM technology- Material Extrusion (MEX). Table 1 summarizes the characteristics of all selected materials, considering the final product is manufactured through MEX technology.

Table 1. Properties of selected materials for MEX technology

Properties	*Units*	*ABS*	*PET*	*HIPS*
Density	Kg/mm^3	1.06E-06	1.54E-06	1.022E-06
Young's Modulus	GPa	2.24	2.758	1.9
Poisson's Ratio	-	0.38	0.417	0.38
Yield Strength	MPa	20	54.4	26.5
Ultimate Tensile Strength	MPa	29.6	55.1	25.5
Thermal Conductivity	W/(mm°C)	1.6E-04	3E-04	1.16E-04
Thermal Expansions Coefficient	J/°C	8.57E-05	7.02E-07	8.4E-05
Specific Heat	J/(kg°C)	1500	2287	1755

To manufacture the main parts of the biped mechanical system, the optimal material is HIPS, as shown in Table 1. The 3D printing equipment used in creating the prototype is a Zortrax M300 plus and the filament is Z-HIPS. The printing parameters were established in correlation with the functional role of the parts. Following the specified guidelines for MEX technology, together with the printing speed and temperature correlated to the material used, the printing location was carefully chosen. This involves reducing the movement of the printing head and getting rid of the passive printing time. A post-processing step was necessary in order to eliminate the redundant parts of the prototype.

4. Sensor integration and measurement of parameters

Figure 6 shows the final biped mechanical system, after it has been modelled, 3D printed and assembled. The system illustrated in Figure 6 also includes the following components: Plusivo board, gas, temperature and motion sensors, a LCD screen, a buzzer, four LEDs (red and green) and resistors with a value of 220 Ω.

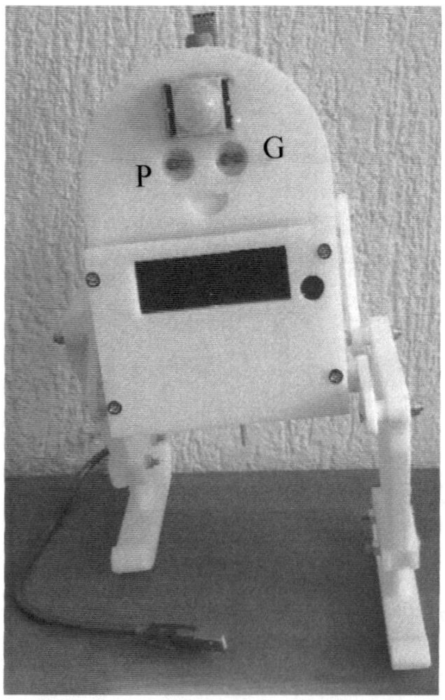

Figure 6. The mechanical system

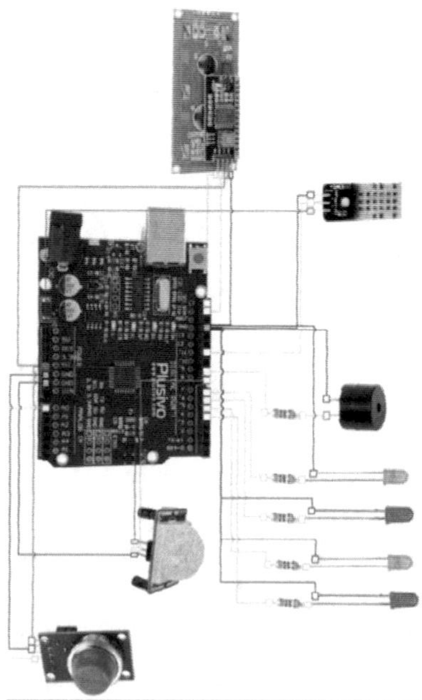

Figure 7. Connection diagram between components

Figure 7 depicts the connection diagram between the electronic components. The DH11 [3,4] sensor monitors the temperature in the ambient environment where the biped mechanical system is located. The values captured by the sensor are displayed on the LCD screen. The connections of this sensor to the Plusivo board is made using the digital pin 11, 5VCC and ground (GND). It can be seen in Figure 8 how the temperature values are shown on the LCD screen. The gas sensor [4] is used to detect certain flammable substances in the air. For the connection of this sensor to the board, three wires are used for an analogic pin, 5VCC and ground (GND). The values generated by this sensor are read from the analog pin A0. If gas leaks are detected in the room, the red LED will light up and the buzzer will give a certain sound. If there are no gas leaks, the green LED will light up and the buzzer will not give any sound.

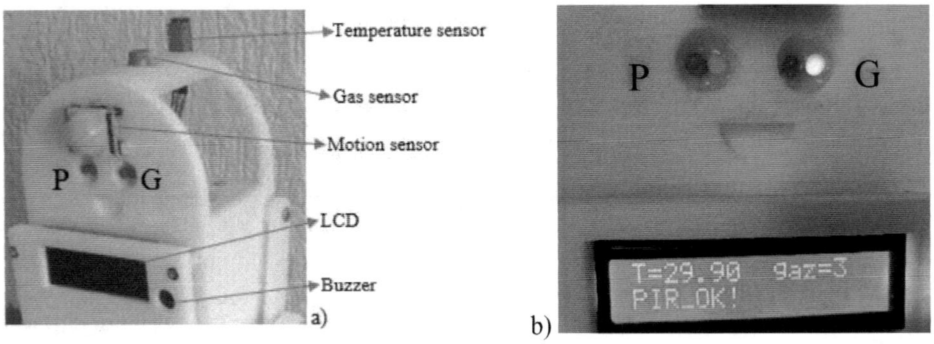

Figure 8. a) The temperature, gas and PIR sensors, b) The value of temperature on LCD screen

Figures 8 and 9 show the working of the system in two different scenarios – with and without the presence of flammable substances in the vicinity (such as lighter gas).
In the first case, when the gas sensor has the proper value, it can be seen the green LED lights up. When the sensor detects certain flammable substances in the air it can be seen the red LED lights up.

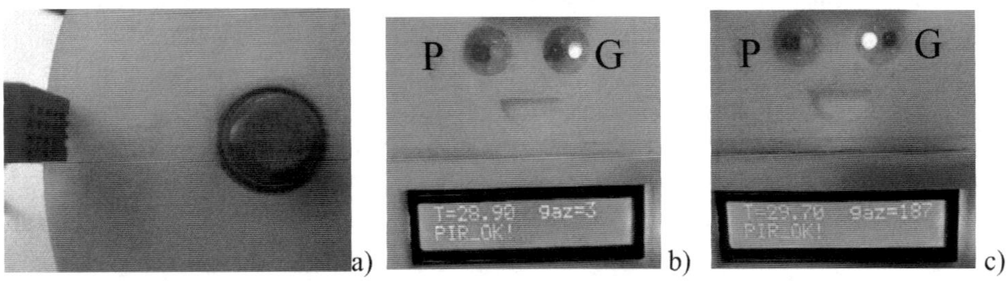

Figure 9. a) The gas sensor: b) normal value for the gas, c) flammable substances in the air

The PIR sensor [3,4] has the role of detecting the presence or absence of movement.

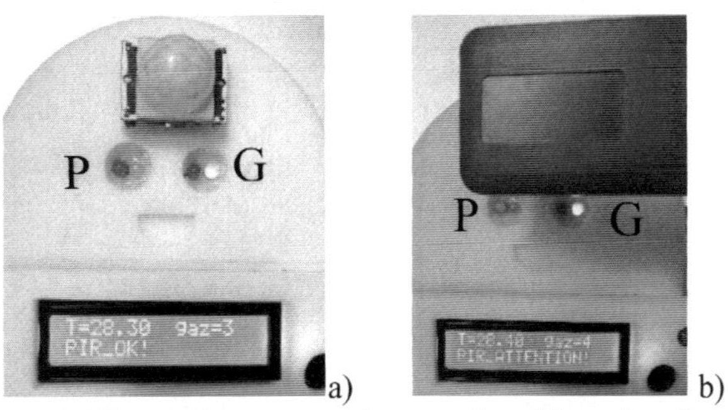

Figure 10. The motion sensor: a) no motion, b) motion detection

The connection of this sensor to the board is made through the digital pin 8, 5VCC and ground (GND). If there is no movement in the detection range of the sensor the green LED will light up, otherwise the green LED will turn off and red LED will light up. In both cases, a message will be displayed on the LCD screen as shown in Figure 10.

5. Conclusions

The biped mechanical system was designed for additive manufacturing and subsequently the sensors and electronics were integrated. In order to obtain a better topology for the prototype, the components went through an iterative development process. Finally, the model's accuracy and functionality was verified after assembling and integrating all the components previously mentioned (after both the 3D printed components and electronic parts were in place). The process of developing the product, with respect to the client's wants and needs, took into consideration the rules specific to 3D print the prototype with Material Extrusion technology, the limitations of the actual equipment, the optimal material and the sizes of the electronic components.

All things considered, the paper's objective—a workable prototype to track environmental parameters—which was outlined in the first chapter was accomplished.

The work presented in this paper, in accordance with previous theoretical and practical work related to AM, goes to show that 3D printing can be a handy tool for rapid prototyping, even when faced with numerous limitations. The inexpensive characteristics of AM both in terms of time and physical resources (materials) makes it a desirable approach especially when coupled with its adaptive and development responsive nature.

6. References

[1] Ulmeanu M., Doicin C., Dezvoltarea Produselor Fabricate Aditiv - Aplicatii ale Analizei Functionale Tehnice, Editura BREN, ISBN 978-606-610-219-3, 427 pages, 2018.

[2] Lehmhus D, Busse M, 2017, Sensor and Electronics Integration in Additive Manufacturing: Technology and Applications. 10.1002/9783527679249.ch9

[3] Bolton, W., 2004, Mechatronics, Electronic Control System in Mechanical and Electrical Engineering, Pearson Studium, München

[4] Cazan-Gheorghiu, C., Electronics and robotics. First steps (in Romanian), Editorial Libris, Brașov, 2018, ISBN 978-606-8953-89-262

[5] Lehmhus D, Aumund-Kopp C, Petzoldt K, Godlinski D, Haberkorn A, Zellmer V, Busse M, 2016, Customized Smartness: A Survey on Links between Additive Manufacturing and Sensor Integration, Procedia Technology, Volume **26**, 284-301, ISSN 2212-0173, https://doi.org/10.1016/j.protcy.2016.08.038.

[6] Gaikwad A, Yavari R, Montazeri M, Cole K, Bian K, Prahalada R, 2020, Toward the digital twin of additive manufacturing: Integrating thermal simulations, sensing, and analytics to detect process faults, IISE Transactions, **52** (11), 1204-1217, DOI: 10.1080/24725854.2019.1701753

[7] Butt J, (2020), "Exploring the Interrelationship between Additive Manufacturing and Industry 4.0" Designs 4, **2** (13), https://doi.org/10.3390/designs4020013

IOP Conf. Series: Materials Science and Engineering 1268 (2022) 012003 doi:10.1088/1757-899X/1268/1/012003

Research concerning ultrasonic chains with torsional vibrations for different industrial applications

D Ghiculescu, R Cretu, O Alupei

Polytechnic University of Bucharest, 313 Splaiul Independentei, Bucharest, Romania

daniel.ghiculescu@upb.ro

Abstract. The paper deals with design, modelling, and testing of an ultrasonic (US) chain with torsional vibrations that includes a tool at its end and could be used for different industrial applications such conventional and nonconventional machining of revolution, helicoidal, and curved surfaces or various types of materials. The US chain includes a horn with helicoidal channel whose profile was studied aiming at growing the amplification of amplitude of the tool, positioned in an antinode point. After the shape design of US horn, finite element analysis was applied to simulate the torsional vibrations of the horn and integrated tool with highest amplitude. Some experiments were made at electrical discharge machining (EDM) aided by torsional ultrasonic vibrations. The results emphasized the technological parameters improvement like machining rate increase and surface roughness decrease, compared to usual EDM without US.

1. Introduction

Ultrasonically (US) aided machining represents a trend of technological hybridization with synergetic effects. It has become very useful nowadays since materials to be machined are characterized by higher resistance and hardness properties, nonhomogeneous structure, e.g., composite materials. When using US assistance of different applications, the main output technological parameters like the machining rate and the surface quality obtained are improved. This is the case of drilling materials used in aeronautic industries, using US longitudinal–torsional vibrations [1]. In the same field, US longitudinal-torsional vibrations were applied at milling Carbon Fiber Reinforced Plastic (CFRP) composites, reducing burrs at entrance and exit of cutting, and reducing cutting forces due to additional effect of fibers shear produced by US [2]. Similar applications were reported at cutting of honeycomb core composites that have poor machinability. Hence, US longitudinal-torsional vibrations were applied reducing burrs and surface quality also by increasing tool speed [3]. The same hybrid mode of US vibration was applied on grinding silicon carbide with similar results [4]. A non-rotative tool associated with the coupling of longitudinal and torsional US vibrations with 20 kHz frequency was used at rock cutting. Thus, it was eliminated the rotative drilling at such materials [5]. In other fields such as surgical applications, US longitudinal and torsional vibrations are combined and are very useful because thus, the cutting energy is focused only on the dissection zone and affects very little the surroundings tissues. A particular application of this principle is represented by an ultrasonic needle used for deep penetrations of bones [6]. It is also mentioned, although this is not the main subject of the paper, that longitudinal-torsional vibrations are applied on large scale for some time at ultrasonic motors [7, 8, 9].

Some constructive solutions regarding different shapes of transversal profile of helicoidal channels were studied to couple longitudinal and torsional vibrations within an ultrasonic chain that uses

IOP Conf. Series: Materials Science and Engineering 1268 (2022) 012003 doi:10.1088/1757-899X/1268/1/012003

Langevin transducers [10]. Another variant to produce US torsional vibrations uses a sandwich transducer with coaxial segments polarized in longitudinal and torsional modes [11].

2. Devices for ultrasonically torsional vibrations aided electrical discharge machining

Torsional US vibrations are very useful to be applied to some complex surfaces that could be generated by electrical discharge machining (EDM) like threads (helicoidal surfaces) and holes with curved axes. Such variants of equipment for specific ultrasonic assistance are the subject of granted patents [12, 13].

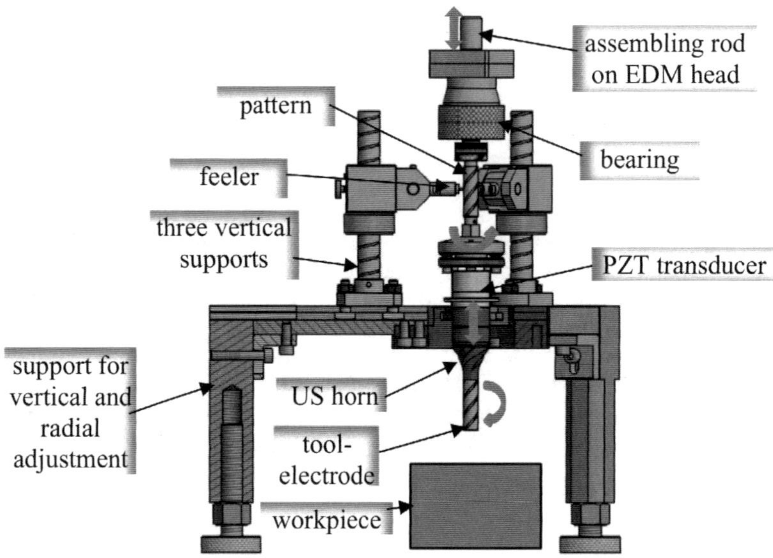

Figure 1. Device for helicoidal surfaces achievement by US torsional vibrations aided EDM [12]

The device that generates helicoidal surfaces by EDM aided by torsional US vibrations comprises the tool-electrode that executes a helicoidal movement due to the pattern and three radial feelers, but at the same time, the tool has US torsional vibrations that improves machining rate and surface quality (figure 1). The US horn that has some helicoidal channels on its exterior surface executes torsional vibrations and the integrated electrode-tool that achieves helicoidal channels on the workpiece. The torsional vibrations are taken from a sandwich transducer that performs longitudinal US vibration being supplied from US generator (not shown).

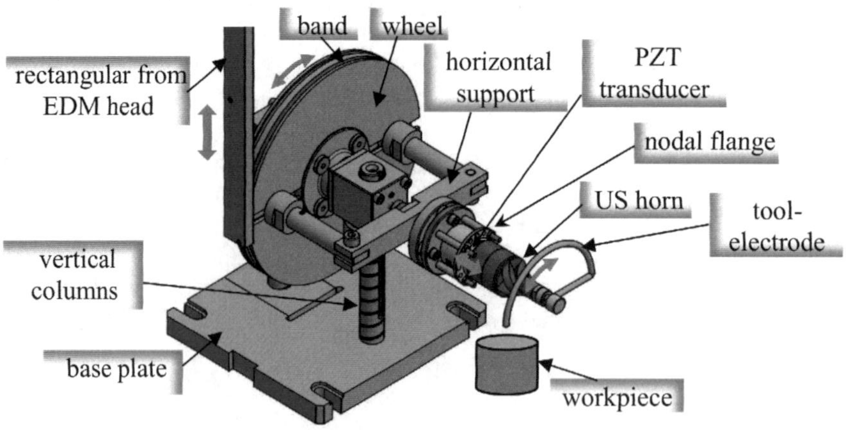

Figure 2. Device for curved (micro) holes achievement by US torsional vibrations aided EDM [13]

IOP Conf. Series: Materials Science and Engineering 1268 (2022) 012003 doi:10.1088/1757-899X/1268/1/012003

The device for curved holes and microholes machining by EDM aided by US torsional vibrations (figure 2) comprises an US chain with a horn with helicoidal channel, integrating a curved tool-electrode that executes torsional vibration transformed from longitudinal vibrations produced by a PZT transducer. The curved tool-electrode also executes rotary movements in both senses needed for EDM, which are taken from translation of a rectangular rod assembled on the EDM head and rotation of a wheel mounted on the rod by metallic bands. The tool-electrode is oriented relative to machined workpiece by vertical columns and horizontal supports.

3. Modeling and simulation of ultrasonic horn

Different shapes (profiles) of US horn sections, the first three taken from [10] and the last one created directly in Comsol Multiphysics, using bult-in tools were studied, and presented in figure 3:

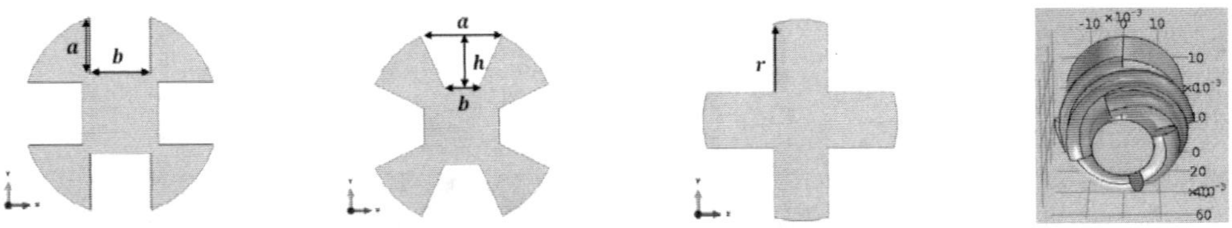

Horn section profile A Horn section profile B Horn section profile C Horn view D

Figure 3. Ultrasonic horns that create US torsional vibrations; A, B, C section profiles, D – horn view

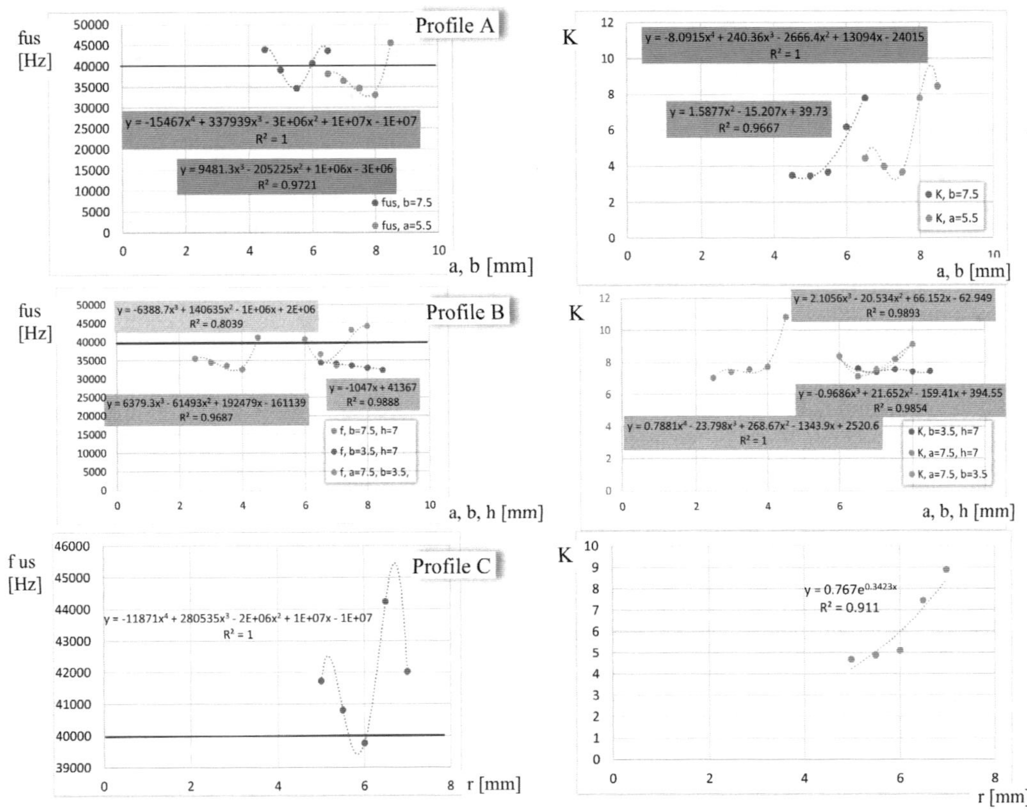

Figure 4. Variation of own frequency (fus) and amplification (K) for different shapes of transversal section of horn profile type A, B, and C

IOP Conf. Series: Materials Science and Engineering 1268 (2022) 012003 doi:10.1088/1757-899X/1268/1/012003

All US horns were parameterized in Eigenfrequency module of Comsol Multiphysics to determine the own frequency and amplification of elongation as ratio between output and input elongations. The corresponding results concerning the variation of own frequencies (fus) and amplifications (K), and its regression functions are presented in figure 4. For profile types A, B, C, the number of turns=1. It can be observed that amplification is growing when the profile parameters, *a, b,* and *r* are increased, i. e. the empty space of the profile grows. It can be found some intermediate values in the studied interval for which the own frequency (i.e. natural frequency or Eigenfrequency) could be overlapped on the nominal frequency of 40 kHz to obtain the resonance frequency (the equality between the own frequencies of the transducer and the horn). To obtain this condition, the overall geometric parameters also contribute, i.e. the length and diameter of the horn.

For horn morphology created in Comsol Multiphysics, different variants were studied, determining the own frequency and amplification, and the most significant are presented in figure 5, together with parametrization of the helixes, where l1= 30.88 [mm] – superior length step, r1= 17.5 [mm]- superior radius step. The material for all variants was steel AISI 4340 with density, 7850 kg/m^3, Young's modulus, $2.05 \cdot 10^{11}$Pa, and Poisson's ratio, 0.28.

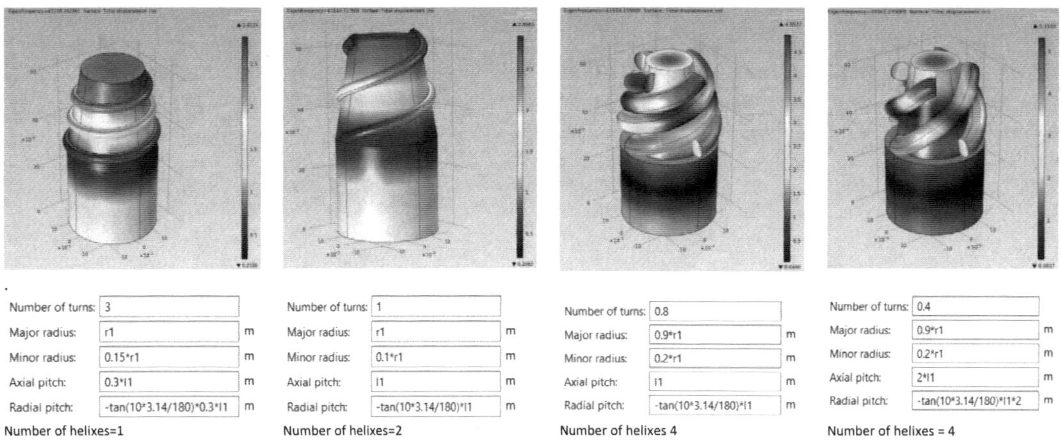

Figure 5. Variation of own frequency (f) and amplification (K) for different constructive variants of helixes added horns

As can be noticed, increasing the number of helixes (*n*) determines the amplification growth. But increasing the *n* parameter moves the maximum amplification (color code and values on right bar) to zones of the helixes, against the lower *n* permits the amplification growing at the tip of the horn. This characteristic becomes important when choosing the type of horn aiming at energy concentration within the zone of interest. For these simulations, the overall dimensions of the horn were the same, l1= 30.88 [mm], r1= 17.5 [mm] as above, l2=32.94 [mm], and r2=10 [mm] for the inferior part of the horn.

The profile horns, type A, B, and C could be machined by CNC milling using a certain shape of the tool, identical with the empty space of the profile in transversal section. The horns from figure 5, could be obtained by additive technologies, e.g., selective laser sintering, taking account of their specific complex shapes. Considering the lowest cost to achieve the US horn for further laboratory tests, respectively the simplest shape of the profile, it was taken the profile A, with the highest amplification and natural frequency as close as possible to the nominal one, of 40 kHz, to be executed on a milling CNC machine.

Besides the profile parameters from above, the Comsol model of US horn was parameterized - figure 6. On this horn, the helicoidal channels were generated in SolidWorks to obtain the geometry of the model. Using Eigenfrequency module from Comsol Multiphysics, the horn own frequency and amplification were obtained further by importing the created geometry from Solid Works – figure 7.

IOP Conf. Series: Materials Science and Engineering 1268 (2022) 012003 doi:10.1088/1757-899X/1268/1/012003

Parameters

Name	Expression	Value	Description
l1	26.367[m...	0.02637 m	length of input step
l2	34.058[m...	0.03406 m	length of output step
d1	45[mm]	0.045 m	diameter of input step
d2	10[mm]	0.01 m	diameter of output step
rr	r1-r2	0.0175 m	fillet radius
ds	10[mm]	0.01 m	diameter of tool
ls	5[mm]	0.005 m	length of tool
r1	d1/2	0.0225 m	radius of input step
r2	d2/2	0.005 m	radius of output step
ych	0.094	0.094	y nodal channel position
rch	0.5[mm]	5.0E-4 m	nodal channel radius
scr	4[mm]	0.004 m	screw radius

Figure 6. Parameters of the US horn

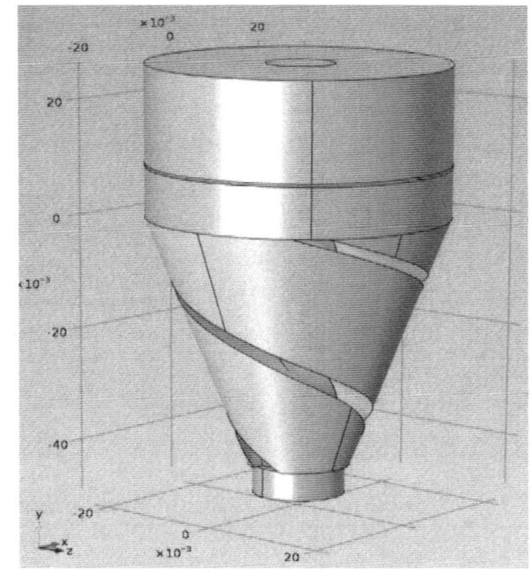

Figure 7. Geometry of the US horn

Since the US torsional vibrations are used at assisting electrical discharge machining, tool-electrode was from copper. In this laboratory stage, for a good transmission of US waves between the horn and the tool, both were made from the same material and one single body. Hence, the following characteristics, the density, 8700 kg/m^3, Young's modulus, $1.532 \cdot 10^{11}$Pa, and Poisson's ratio, 0.32, for this material, were introduced in the material node of the Model Builder.

For boundary conditions, to determine the own frequency were set to free. The mesh made from tetrahedral elements and its characteristics are presented in figure 8. The results concerning the eigenfrequency and amplification (color code on amplification bar below) of the designed horn are presented in figure 9.

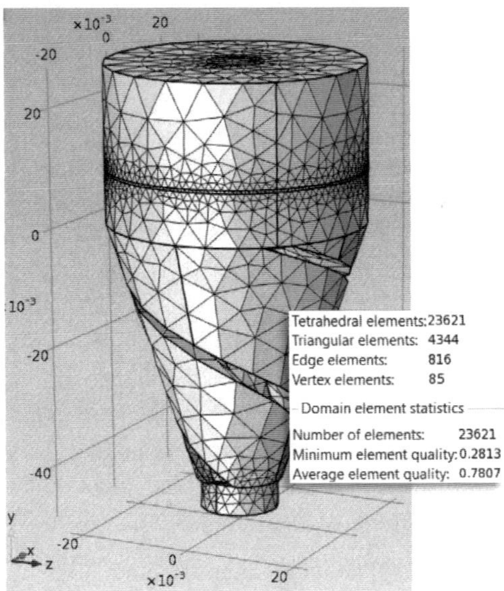

Figure 8. The mesh characteristics

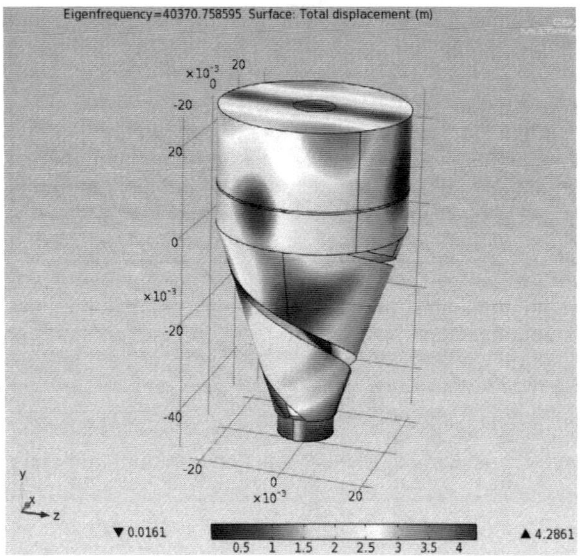

Figure 9. The eigenfrequency and amplification of the US horn

This shows the own frequency of the horn, which determines by simulation only torsional vibration, close to the nominal frequency of 40 kHz, which denotes the possibility of obtaining the resonance condition, the same with transducer one.

4. Experiments of electrical discharge machining aided by US torsional vibrations

The US horn type A with helicoidal channels was achieved on CNC lathe, MAZAK INTEGREX 300Y by importing the CAD file created in Solid Works in CAM software installed on the machine. The contact surface with the transducer was polished at surface roughness, Ra=0.2 µm, and flatness deviation was less than 1 µm. The frontal contact surfaces on the transducer were diamond burnished with similar characteristics [14]. Then, the US horn was assembled with the transducer using a stud.

The experimental determination of own frequency of the assembled ultrasonic chain effectuated on the experimental stand (figure 10). The end of ultrasonic chain, respectively the tool-electrode was immersed into water. Thus, the own frequency of ultrasonic chain was determined as 40.75 kHz, using a tone generator, close to that obtained by Comsol simulation (figure 11). Then the supply frequency of US generator dedicated to assisted microEDM (with adaptive control) was adjusted for this initial resonance frequency. The cavitation bubbles were observed within the water vessel at this frequency.

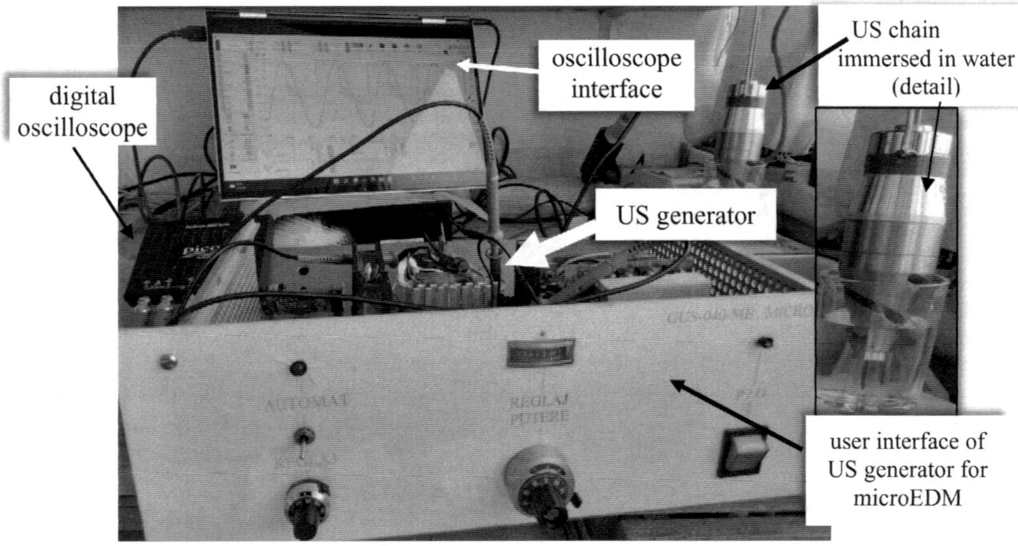

Figure 10. Resonance frequency determination and supply frequency adjustment – test stand

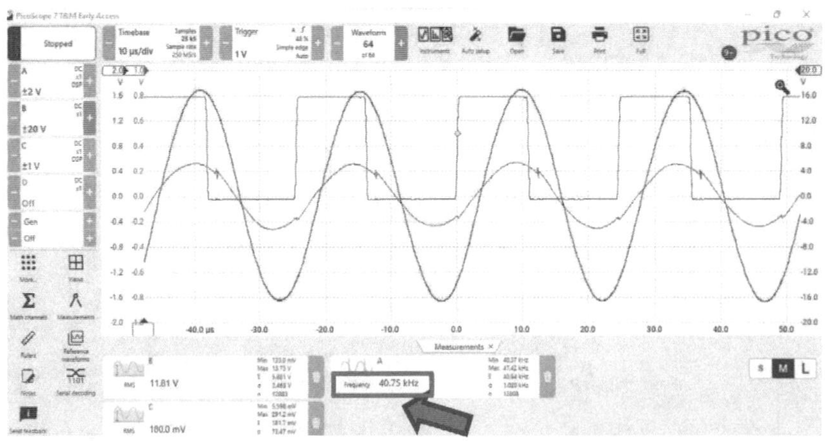

Figure 11. The ultrasonic chain own frequency of 40.75 determined on digital oscilloscope

Some surfaces on a sample from X210Cr12 steel, thermally hardened at 60 HRC, were machined comparatively by usual EDM and US aided (torsional vibrations) EDM, with the equipment of US assistance, containing this US chain, mounted on Romanian machine ELER 01 – figure 11.

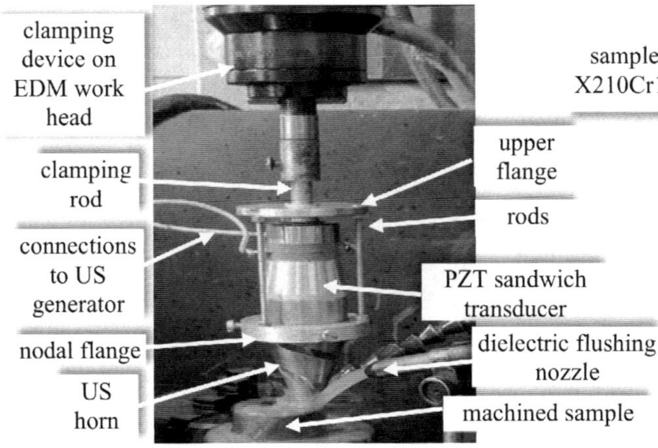

Figure 12. The equipment of US torsional vibrations on ELER 01

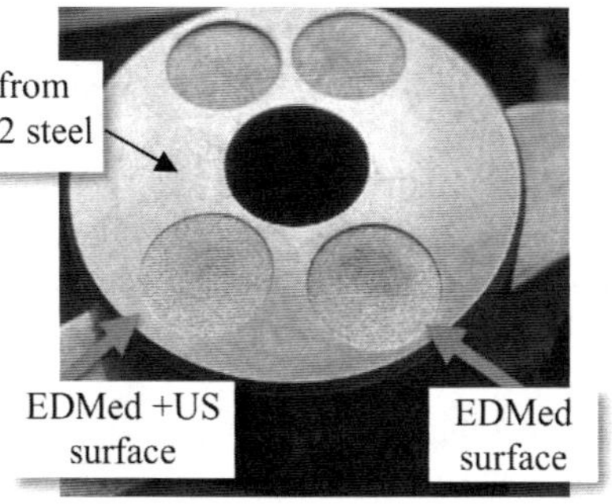

Figure 13. Surfaces machined by EDM and EDM+US-torsional vibrations

The machined surfaces were evaluated in terms of depth, for machining rate (Vw) and roughness (Ra), primarily using a Charmilles ruler and finally by a surface roughness tester Taylor Hobson Surtronic. The results are presented in figure 14.

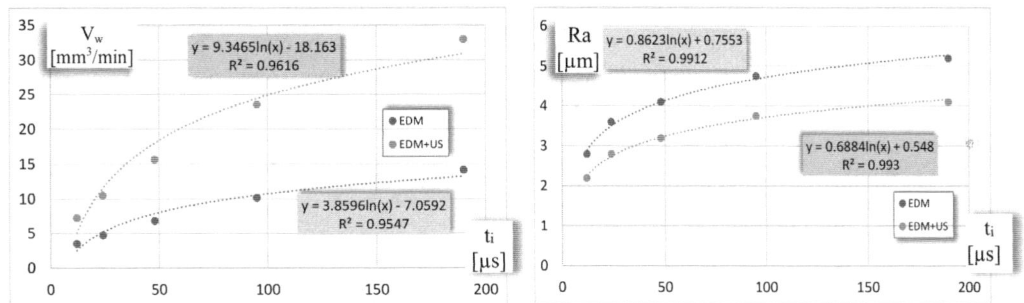

Figure 14. The variation of machining rate (V_w) and surface roughness (Ra) at EDM and EDM+US-torsional vibrations depending on pulse time (t_i)

As it can be noticed, the increase of machining rate at EDM+US (torsional vibrations) compared to usual EDM is at least 200% with higher values at longer pulse time. This is based on higher probability to overlap the pulse time on cumulative microjets stage, the collective implosion of the gas bubbles from the gap at the final of dielectric liquid stretching semiperiod [15]. The roughness decreased due to ultrasonic assistance with at least 20%. This could be based on discharge energy dissipation because of torsion movement of the frontal surface of the tool-electrode. Moreover, this could be the effect of shock waves produced by US cavitation that are oriented parallel to machined surface [15]. This can produce additional shear load to remove the micropeaks, and this reduces the machined surface roughness.

5. Conclusions

All these improvements reported at ultrasonically longitudinal-torsional vibrations aided mechanical machining are based on reduction of friction coefficient due to the intermittent contact between the tool and the workpiece produced by ultrasonic vibration, and thus, the cutting forces decrease. Ultrasonic torsional vibrations applied to EDM can significantly increase machining rate and surface roughness. The results presented in this study are applied on flat surfaces, but torsional vibrations are more appropriate to revolution surfaces like helicoidal and even curving holes.

Some disadvantages of US longitudinal-torsional vibrations consist in relatively close resonance frequency of different vibrations modes that requires a very strictly adaptive control of US generator. During applications performed various perturbations like working temperature of US chain and tool wear could modify the resonance frequency. Likewise, the very intricate design and execution of US horn could prolong the time of fabrication preparation.

Further research will be approached in connection to technological performances at the ultrasonically aided threading and curving holes machining by EDM, using conjugated tools assembled at horn end.

Acknowledgments
Authors wish to acknowledge the assistance from Eng. Petre Francisc at CNC machining of ultrasonic horn, and Eng. Stefan Nitisor at US generator adjustment and own frequency determination.

References

[1] Amini S, Soleimani M, Paktinat H and Lotfi M 2017 Effect of longitudinal–torsional vibration in ultrasonic-assisted drilling *Mat. Manuf. Proc.* 32:6 pp 616-622
[2] Liu J, Chen G, Ren C, Qin X, Zou Y, Ge J, 2020 Effects of axial and longitudinal-torsional vibration on fiber removal in ultrasonic vibration helical milling of CFRP composites *J. Manuf. Proc.* **58** pp 868-883
[3] Xiang D, Wu B, Yao Y, Liu Z and Feng H 2019 Ultrasonic longitudinal-torsional vibration-assisted cutting of Nomex® honeycomb-core composites *Int. J. Adv. Manuf. Technol.* 100, pp1521–1530
[4] Chen Y, Su H, He J, Qian N, Gu J, Xu J, Ding K, 2021The Effect of Torsional Vibration in Longitudinal–Torsional Coupled Ultrasonic Vibration-Assisted Grinding of Silicon Carbide Ceramics *Materials* 14, 688
[5] Harkness P, Cardoni A and Lucas M, 2008 Vibration considerations in the design of an ultrasonic driller/corer for planetary rock sampling *Proc. Of ISMA 2008: Int. Conf. On Noise And Vibration Engineering* (Leuven), pp 3477-3486
[6] Cleary R, Wallace R, Simpson H, Kontorinis G, Lucas M 2022 A longitudinal-torsional mode ultrasonic needle for deep penetration into bone *Ultrasonics* **124** 106756
[7] Tsujino J 1998 *Smart Mater. Struct.* **7** p 345
[8] Tomikawa Y, Adachi K, Aoyagi M, Sagae T and Takano T 1992 Some constructions and characteristics of rod-type piezoelectric ultrasonic motors using longitudinal and torsional vibrations, *IEEE Trans. on Ultrasonics, Ferroelectrics, and Frequency Control* **39** 5 pp 600-608
[9] Yi Y, Gausmann R, Seemann W and Zhong J A 2005 Note on a Longitudinal and Torsional Type of Ultrasonic Motor with Two Stators, *J. of App. Math. Mech.* **5** pp 147-148
[10] Al-Budairi H D 2012 Design and analysis of ultrasonic horns operating in longitudinal and torsional vibration *PhD thesis* University of Glasgow
[11] Shuyu L 1997 Sandwiched piezoelectric ultrasonic transducers of longitudinal-torsional compound vibrational modes, IEEE Trans. on Ultrasonics, Ferroelectrics, and Frequency Control **44** 6 pp 1189-1197
[12] Marinescu N I, Ghiculescu D, Alupei O, Popa L 2021 Device for helicoidal channels achievement by ultrasonically aided electrical discharge machining, on exterior and interior surfaces of electrically conductive workpieces, Romanian patent RO-0013148
[13] Ghiculescu D, Marinescu N I, Alupei O, Popa L 2021 Device for curved holes and microholes achievement by ultrasonically aided electrical discharge machining, Romanian patent RO-00131489
[14] Ferencsik V, Varga G 2020 Examination of Surface State-change on diamond burnished aluminium components, *Proc. Intern. Symp. for Production Research* (Vienna) **47** pp. 535-544
[15] Ghiculescu D, Marinescu N I, Alupei O 2016 On overlapping the pulses on cumulative microjets stage at ultrasonically aided electrical discharge machining, *Appl. Mech. Mat.* **834** pp126-131

IOP Conf. Series: Materials Science and Engineering 1268 (2022) 012004 doi:10.1088/1757-899X/1268/1/012004

Prescribed precision for medical products obtained with additive manufacturing technologies

N Ionescu[1], M E Ulmeanu[1], C V Doicin[1], C Ionescu[2] and P Spânu[1]

[1] Manufacturing Department, Faculty of Industrial Engineering and Robotics, University POLITEHNICA of Bucharest, Splaiul Independenței 313, Sector 6 Bucharest, Romania
[2] GHI Hornos Industriales S.L., Barrio de Aperribai, 4 – 48960 Galdakao, Biscay, Spain

mihaela.ulmeanu@upb.ro

Abstract. The paper highlights the importance of corelating the function of a medical product, with its' design and manufacture stages within the global product design and development process. Dimensional accuracy, macro-geometric and micro-geometric shape accuracy, relative surface position accuracy and other prescribed characteristics of medical products should take into consideration a variety of factors, such as custom requirements, regulated by the product functions. In an ideal scenario, with unlimited resources, prescribed characteristics dictate the selection of the manufacturing technology and equipment. Nonetheless, in most cases the design stage takes into consideration capabilities of available technology, correlated with bespoke requirements. In this context, the authors propose a clear distinction between precision, accuracy, repeatability, and resolution, as to properly address each issue in the lifecycle of a medical product manufactured using additive technologies. Using a real application for an upper limb prosthesis, the nominal position precision is prescribed, based on size-dependent geometric tolerances, for a cylindrical surface in relation to three reference surfaces, with the application of the maximum and minimum material principles.

1. Introduction

Nowadays, the design and manufacture of medical products is in an accelerated development due to cutting-edge technological developments, one of the most important is the exponential increase in the performance of additive technologies [1, 2]. The functional role of these medical products requires the establishment, during the design stage, of specific characteristics regarding dimensional accuracy, macro-geometric and micro-geometric shape accuracy, relative surface position accuracy, etc. In establishing these characteristics, the functional role of the surfaces must be taken into account, i.e. the fact that some surfaces are in contact with mucosae or external tissues (skin), others are assembly surfaces and other surfaces are external surfaces that must have dimensions and anatomical shapes and a corresponding appearance as close as possible to the natural human one. Additive manufacturing (AM) is becoming a preferred technology for the manufacture of bespoke medical products, thus the functionality of the product and specific technology capabilities influence the design process.

If we consider the product design and development process [3] proposed by Ulrich and Eppinger and applied by a variety of other specialists, a product goes through a sequence of steps or activities such as design, manufacture and market [4]. The sequence begins with the identification of a need (market opportunities) and ends with the sale or delivery of the product to the customer. The process is iterative, multidisciplinary and allows the "transformation" of a need-based idea identified in the market into a successful product. In this scenario, the design process of a medical product manufactured using AM technologies should adhere to bespoke overall functions, individual surface functions and machine capabilities. Prescribed precision of AM medical products is somewhat similar to traditional manufactured products and is highly dependent on product requirements, which later construct final product specifications. Final medical product specifications might be difficult to achieve due to a variety of trade-offs which need to balance the functions' satisfaction degree. Typically, there could be trade-offs between a product's weight and cost, its quality and cost, or its cost and productivity [5]. For example, the smaller the overall dimensions of a memory device (microSD vs external HDD), the higher the cost, taking into consideration the same technical specifications. Ideally, clients desire high-quality goods that are produced affordably and quickly. The reality is that high-quality goods cannot be produced quickly at a low cost. Customers also want their goods to be customized, thus enabling the development of innovative production methods so as to make these personalized items possible, but at affordable prices. Due to the personalization characteristic of medical products, quality usually overpowers cost, thus higher precisions are prescribed to ensure customer compliance and reliability, within the working parameters of the medical application. Hereafter, we discuss on the main AM capabilities and prescribed precision for medical products and how they relate to the development stages of a product.

2. Manufacturing capabilities and prescribed precision of medical products

For the manufacture of medical products, technologies, methods and procedures must be selected that ensure a processing accuracy corresponding to the prescribed precision. Thus, if additive manufacturing (AM) is selected as the main manufacturing technology, depending on the process used, subsequent machining processes such as CNC milling may or may not be necessary [6]. It important to make a clear distinction between precision, accuracy, repeatability, and resolution, as to properly address each issue in the lifecycle of a medical product.

Precision describes how close measured values are to each other [7]. Precision refers to the fabricated physical product. Accuracy describes how closely a manufacturing machine's output conforms to a tolerance within a specified dimensional range. In this case, how close a 3D printer's measured value is to the prescribed precision of medical devices like implants, prosthesis and orthoses. Repeatability captures the 3D equipment's capability to produce consistent output, throughout a defined period of time. Manufactured results should be similar within the tolerance range. Resolution refers to the smallest measurement a used 3D printer can reproduce. Accuracy, repeatability and resolution refer to the 3D equipment and its' capabilities [8, 9]. We must consider that when dealing with AM, accuracy, repeatability and resolution are highly dependent on the technology and often on the individual machine that is being used to manufacture the final product.

These concepts are widely known and used by designers and manufacturing engineers. Nevertheless, there is a common misconception regarding how these concepts apply in the AM field. Currently there are professional groups that use resolution characteristics, like layer thickness or dots per inch, in order to address accuracy aspects. Resolution does not, in fact, translate directly to a system's overall accuracy, and it is specifically important to take notice of these nuances when dealing with AM equipment. Some 3D printing machines manufacture parts with fine features but lower overall dimensional accuracy, whilst other machines manufacture parts with lower, but acceptable, resolution and excellent accuracy [1].

The difference in accuracy and resolution did not matter all that much when AM was still in its early stages and employed mostly for model construction and visual prototypes [9]. Technologies were used to manufacture prototypes rapidly and it was enough for a 3D printing machine to turn out an approximate overall dimension part, as long as it met the final aesthetic goals. Nowadays, AM systems have a lot more requirements and custom specifications to meet. It is expected to consistently produce working prototypes, fixtures, or finished parts that must ensure the same exact precision and repeatability demands as traditional manufacturing procedures like CNC machining, injection molding, and casting [1, 2, 4].

Considering an unlimited resource scenario, the authors propose the usage of the above concepts in accordance with a products' life cycle, as defined in table 1.

Table 1. Product design and development in an unlimited resource scenario

PRODUCT DESIGN AND DEVELOPMENT	PLANNING	Need ↓ Function ↓	Custom requirements
	DEVELOPMENT	General & Detail product design	Product specifications - prescribed precision
	PRODUCTION	↓ Additive Manufacturing ↓ Validation ↓	Equipment & Technology capabilities - accuracy - repeatability - resolution
	USAGE	Usage & Maintenance ↓ Decommissioning	Recalibration Recyclability & Reusage

In a limited resource scenario, the production and development stages go into a feedback loop between what is required and what is possible to obtain, given the available resources.

The incremental way that AM equipment manufactures parts from CAD models, gives way to confusion between accuracy and resolution. It is a common misconception to consider a systems' accuracy equal to the height of the deposed layer and the width of the smallest feature, thus considering that a fine resolution translates automatically into an accurate part [10]. Depending on the application, this is not necessarily true. In correspondence to accuracy, resolution is significant only when manufacturing parts with very small functional features. Thus, when an application requires, through its' need and functions, micro and nano functional features or wall thicknesses, the resolution can impact a 3D system's ability to accurately build the small features. Nevertheless, this is not the case of general-purpose medical products like implants, prostheses and orthoses. The medical industry commonly uses AM machines to build parts which are several dozen millimeters in overall dimensions and have tolerance capabilities no tighter than several thousandths of a millimeter. At this scale, regardless of the method of deposition, the resolution of the 3D machine does not control the overall accuracy of the medical part or the repeatability of the process.

As low-cost medical products are more frequently manufactured using material extrusion, Fused Deposition Modelling (FDM) is one of the most commonly selected technology in these cases [2]. Professional FDM machines can manufacture parts with layer resolutions down to 0.1 mm, turning out components that are not as smooth as injection-molded ones. Nonetheless, they have an overall accuracy or tolerance of ± 0.08 mm, which comfortably sets within the average tolerance for an injected molded part, which is around 0.01 mm. A high-resolution 3D printing system is needed only when the product requirements ask for high-end surface finish or sharp edges. It is important to take into consideration that a high surface roughness does not imply accurate parts. Accuracy is in fact dependent on a 3D printing system's capacity to control the motion of the material addition mechanism across the full build volume, regardless of the surface finish of the final part [10].

As FDM systems are more frequently used for low-cost and low-volume bespoke medical products, accurate overall dimensional accuracy is key for these manufactured products in order for them to function in accordance with requirements and still have a high degree of bespoke patient compliance.

Moreover, as AM products are designed with more demanding functional requirements, their accuracy tolerances are prescribed similarly as those of conventional manufacturing processes. Tolerances are prescribed in hundredths of a millimeter over given part dimensions, and not dots per inch or layer height, as would be accustomed in AM practice [11].

The accuracy of a manufactured component is only one of three important key factors to take into account when dealing with AM products. Repeatability and stability of dimensions over a period of time are also important in obtaining an adequate part. If we consider bespoke medical applications, like implants, prostheses and orthoses repeatability is not as significant as long-term stability. Part accuracy is maintained over a given period of time primarily through material stability, which is influenced by a variety of factors, both external and internal to the used AM process. Components may contract, expand, or distort as a result of environmental factors such heat, moisture, and UV exposure as well as remnant stressors from various AM techniques [2, 4].

The function of the product usually dictates the type of material and, consequently the additive technology for its' manufacture. For example, a dental surgical guide can be manufactured from resin using Digital Light Processing (DLP), due to its high accuracy output. Nonetheless, photopolymers are known to be less dimensionally stable over time than thermoplastics. Thus, the surgical guide will change shape and dimensions over time. Considering the application, DLP is still the right choice for this medical product, as surgical guides are used almost always within a few days of manufacturing and are single use for a specific patient. In this case, the function of the product enables the benefit of the high accuracy and disregards the downside of low material stability. On the other hand, industrial thermoplastic additive components manufactured of acrylonitrile butadiene styrene (ABS) or polycarbonate (PC) do not display these alterations [12], making them suitable for manufacturing of custom long-term prostheses and orthoses. The same discussion is in place for metal-based AM systems used for manufacturing of medical products.

When considering accuracy and precision in AM of medical devices, there are many different aspects to take into account, especially: AM technology; Materials; Post-processing; Ecosystem and Calibration. As discussed above, accuracy and precision are connected and complementary depending on the application, but one should consider that 3D printer specifications alone do not represent final dimensional accuracy. Nonetheless, the most typical 3D printing machine tolerances for the most common plastic 3D printing processes are shortly covered hereafter.

FDM allows a ± 0.5% (lower limit ± 0.5 mm) tolerance range when printing layers of thermoplastic filaments are extruded by a nozzle [13]. The technology lacks the precision and capacity for complex details that other AM technologies can provide. High-end FDM systems address other disadvantages such as warping and shrinkage, which tend to appear because of internal stress as the printed part cools down at different rates.

Powder bed fusion (PBF) technologies, such as Selective Laser Sintering (SLS) and Multi Jet Fusion (MJF) allow a tolerance of ± 0.3%, with a lower limit of ± 0.3 mm [14]. PBF technologies are ideal for manufacturing of complex geometries, including interior features, undercuts, thin walls, and negative features, because there is no need for support structures, as the unfused powder supports the part during printing. This technology characteristic, combined with long-term stable materials make it perfect for bespoke medical products like orthoses.

Vat processes like Stereolithography (SLA) and Digital Light Processing (DLP) offer a tolerance range of ± 0.2%, with a lower limit of ± 0.1 mm [15]. Without minimizing the main disadvantage (low part stability over time) these processes can achieve fine details and can consistently produce high quality results. Support structures may be necessary for resin 3D printed medical products to achieve dimensional accuracy, particularly when the model geometry calls for complicated geometries or big, thin walls, which is the case for some anatomical models.

Mechanical characteristics of the selected materials can also have an impact on accuracy by influencing how likely a print is to distort over time [1, 4]. For resin-based processes it is aimed to choose materials with high before post-curing modulus in order to maximize the chance of 3D printing very thin parts with high precision. For material extrusion processes the higher the extrusion

IOP Conf. Series: Materials Science and Engineering 1268 (2022) 012004 doi:10.1088/1757-899X/1268/1/012004

temperature, the more prone to warping the final part is. In this case, PLA is not as susceptible to warping as ABS, although it has inferior strength characteristics.

Usually, products that are manufactured using additive technologies require some sort of post-processing steps. Some of these can influence the dimensions and the surface of the parts, which in turn influence accuracy and tolerances. They are often taken into consideration during the design and print preparation stages, but some might require specific implementations in accordance with product functions. Post-processing steps can entail the usage of traditional manufacturing processes (milling, drilling, cutting, sanding etc.), which use standard practice accuracy, repeatability, precision and tolerances and are provided for within the product design and development overall process.

Lastly, when manufacturing accurate and precise 3D printed medical products it is important to pay attention to more than the machine and consider the entire process. The success of the final result (achieving the prescribed precision) is down to a variety of factors related to the ecosystem and calibration of the 3D machine: print preparation software; printing technology; quality and calibration of the printer and its components; quality of 3D printing materials; post-processing tools and methods; overall manufacturing environment (vibrations, temperature, humidity etc.).

It is ultimately down to the knowledge and experience of the product designer to select and decide on the perfect combination between equipment and technology capabilities in order to achieve the prescribed product precision.

3. Prescribing nominal position precision based on size-dependent geometric tolerances

The nominal position is a feature of the surface position often used in the design work, by which it is required that the tolerance to the relative position of certain surfaces is not determined by the tolerances of the nominal dimensions of the relative position [16], but to have an imposed value, in general, for reasons related to the reduction of manufacturing costs.

If precision is prescribed at the nominal position, the nominal dimensions that determine the relative position of the surfaces to the reference surfaces (datums) become "theoretically exact dimensions" [16] (or basic dimensions in ASME [17]). One of the cases often used in design activity, highlighted in both ISO and ASME standards [17, 18], is the prescription of the precision at the nominal position of a cylindrical surface (hole) with respect to three reference surfaces (Figure 1), with or without the application of the maximum material principle.

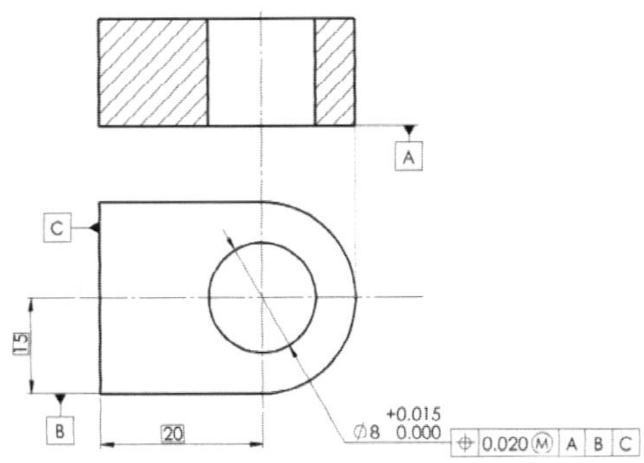

Figure 1. Example of prescribing nominal position precision

To demonstrate the process an AM upper limb prosthesis is considered as a case study (Figure 2). The nominal position precision is prescribed, based on size-dependent geometric tolerances, for the cylindrical surfaces in relation to three reference surfaces, with the application of the maximum and minimum material principles. As a reference, the cylindrical surface with a designed diameter of ∅8 mm from the phalanx articulation is selected for this study. It must be noted that the nominal position precision is prescribed similarly for all cylindrical articulation from the construction of this custom prosthesis.

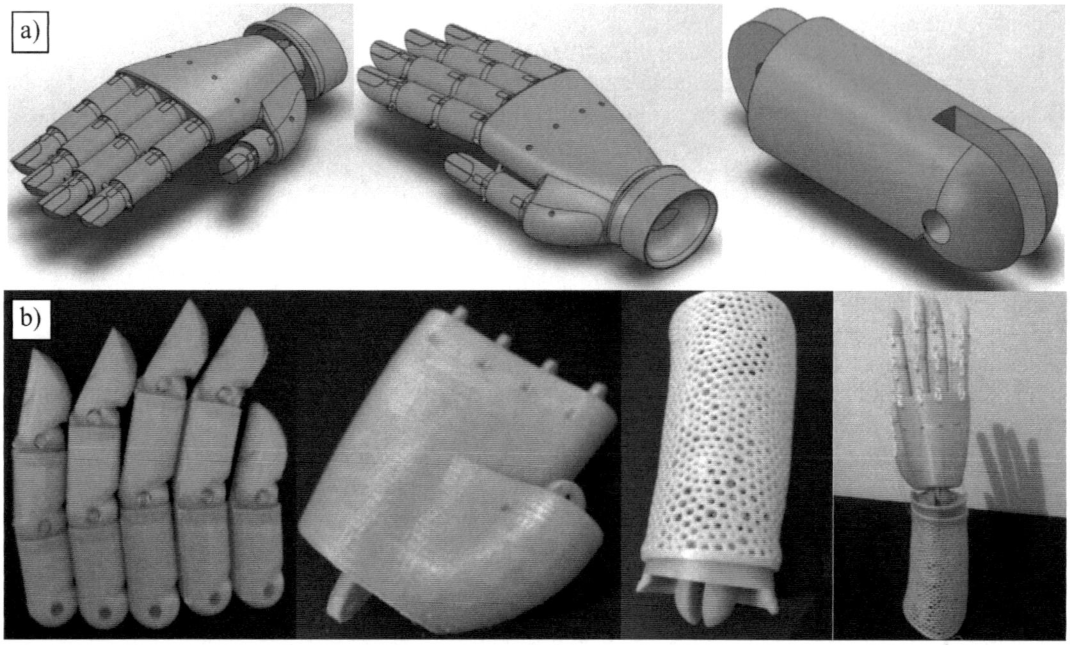

Figure 2. Upper limb prosthesis: a) CAD model of palm and phalanx; b) 3D printed components and assembled prototype [19]

Considering the cylindrical surface of the phalanx, the interpretation is as follows: the effective axis of the toleranced surface can float inside a cylindrical tolerance zone with the axis located at the theoretically exact dimension of 20 mm from the reference surface (datum) C and at the theoretically exact dimension of 15 mm from the reference surface (datum) B, perpendicular on the reference surface (datum) A. If the prescription is based on the principle of the maximum material, the diameter of the tolerance zone varies between 0.020 mm at the maximum material (∅ 8.000) to 0.035 = 0.020 + 0.015 mm at the minimum material (∅ 8.015).

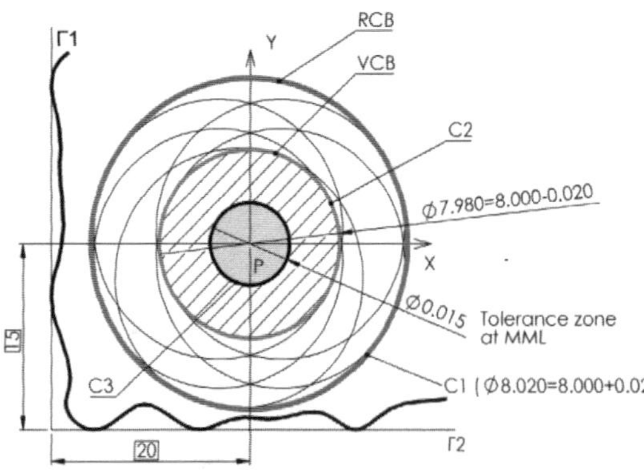

Figure 3. VCB and RCB generation at maximum material

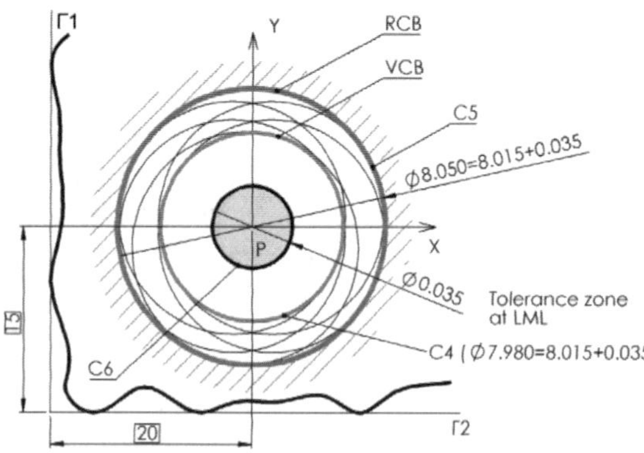

Figure 4. VCB and RCB generation at minimum material

The schemes shown in figures 3 and 4 are particularly important for the design of the phalanx and can be further explained as follows.
To the maximum of material:

- The hole has an actual size of Ø 8.000;
- The axis of the cylindrical tolerance zone is placed 20 mm from the datum plane established by datum feature C, 15 mm from the datum plane established by datum feature B and perpendicular to the datum plane established by datum feature A;
- The diameter of the tolerance zone is Ø 0.020 mm;
- The geometric locus of the points farthest from the theoretically exact position of the axis of the tolerance zone when the actual axis of the hole floats in the tolerance zone is a cylinder with a diameter of Ø 8.020 = 8.000 + 0.020; this cylinder called Resultant Condition Boundary - RCB in standards [16, 17, 18, 20];
- The geometric locus of the points closest to the theoretically exact position of the axis of the tolerance zone when the actual axis of the hole floats in the tolerance zone is a cylinder with a diameter of Ø 7.980 = 8.000 - 0.020; this cylinder called in standards Virtual Condition Boundary - VCB "which must not be violated" according to ISO and ASME [17, 18].

To the minimum of material:

- The hole has an actual size of Ø 8.015;
- The axis of the cylindrical tolerance zone is placed 20 mm from the datum plane established by datum feature C, 15 mm from the datum plane established by datum feature B and perpendicular to the datum plane established by datum feature A;
- The diameter of the tolerance zone is Ø 0.035 = 0.020 + 0.015 mm;
- The geometric locus of the points farthest from the theoretically exact position of the axis of the tolerance zone when the actual axis of the hole floats in the tolerance zone is a cylinder with a diameter of Ø 8.050 = 8.015 + 0.035 (RCB);
- The geometric locus of the points closest to the theoretically exact position of the axis of the tolerance zone when the actual axis of the hole floats in the tolerance zone is a cylinder with a diameter of Ø 7.980 = 8.015 - 0.035 (VCB), the same value as that corresponding to the maximum material and, also identical to that which would result for any intermediate position, for actual hole diameter values between Ø 8.000 and Ø 8.015 mm.

3.1. LabView validation for the prescription of nominal position accuracy based on size-dependent geometric tolerances

Previous work of the authors include the development of a LabView programming environment to help design engineers and technologists work with the prescribed nominal position based on the maximum material principle [19]. Generalizing the previous problem, the equations of all cylinders (circles, in each section normal to the axis of the hole) were written that intervene in the explanations in figures 3 and 4, considering the case of a certain bore with the diameter $\boldsymbol{\varnothing\, D_{EI}^{ES}}$ and for a certain tolerance at the nominal position T_i (figure 5).

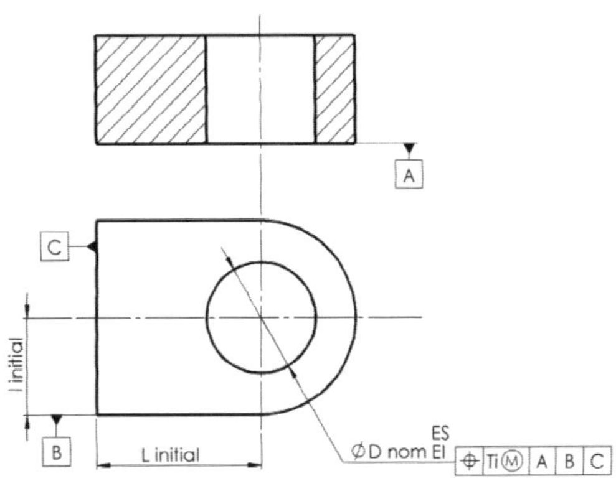

Figure 5. The general case of prescribing nominal position accuracy for a hole

The application [19] allows the user to enter all the input data and generates the three cylinders: tolerance zone, VCB and RCB (figure 6). The theoretically exact dimensions ($L_{initial}$ and $l_{initial}$) are not tolerated directly but, implicitly, their tolerances result as symmetrical deviations from the nominal position tolerance ($L_{calculated}$ and $l_{calculated}$).

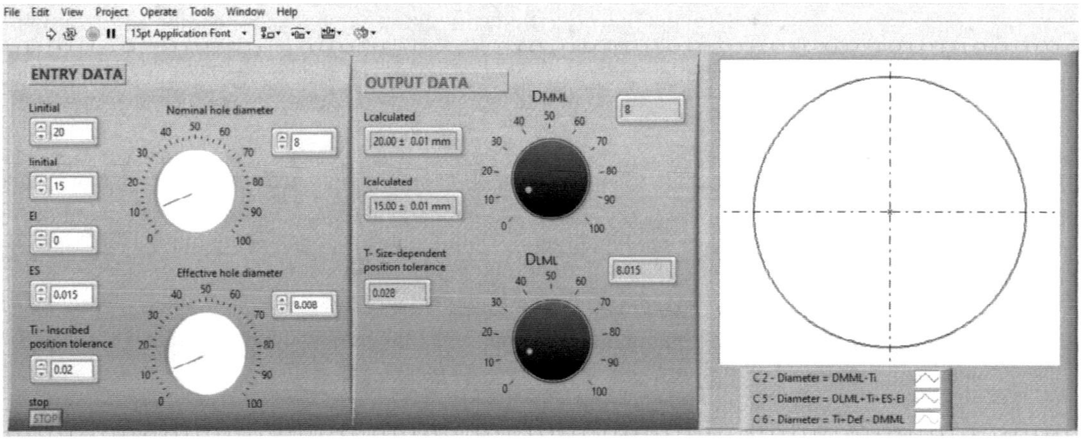

Figure 6. Front panel of the virtual instrument – medical case study validation

4. Conclusions

The paper discusses on the importance of corelating the function of a medical product, with its' design and manufacture stages within the global product design and development process. Together with the function of the product, custom requirements usually dictate the dimensional accuracy, macro-geometric and micro-geometric shape accuracy, relative surface position accuracy and other prescribed characteristics of medical products. AM products are designed with more demanding functional requirements, thus their accuracy tolerances are prescribed similarly as those of conventional manufacturing processes. Stability of overall dimensions over a period of time is also important in obtaining an adequate bespoke part in the medical industry, compared with repeatability of the results, as products are often customised on patient anatomical data. Resultant Condition 3. When prescribing nominal position precision based on size-dependent geometric tolerances, Boundary and Virtual Condition Boundary have been graphically identified for both maximum and minimum material principles in a cylindrical surface with an actual size of ∅ 8.000 mm.

A software application has been used to validate the selected case study and demonstrate the proposed principles of minimum and maximum of material for prescribing the nominal position precision based on size-dependent geometric tolerances for a upper limb prosthesis. The software applications assist design engineers and technologists in interpreting position accuracy prescriptions based on individual size-dependent tolerances, such as those related to nominal position, being very useful for example in ensuring the minimum thickness of some walls. A similar application can be adapted for external surfaces, with the mention that in this case the maximum material diameter is D_{MML} = d_{max} and the minimum material diameter is d_{min} D_{LML}. Also, similar applications are in progress for the prescription of the precision at the nominal position based on the minimum material principle, in which case the listed tolerance is valid at $d_{LML} = d_{min}$ for shafts and $D_{LML} = D_{max}$ for holes and the difference between the effective diameter and the diameter at minimum material is added to the nominal position tolerance.

References

[1] Prakash C, Singh S and Ramakrishna S 2022 *Additive, Subtractive, and Hybrid Technologies* (Springer Cham)
[2] Banga H K, Kumar R, Kalra P and Belokar R M 2022 *Additive Manufacturing with Medical*

Applications (CRC Press)
[3] Ulrich K T and Eppinger S D 2012 *Product Design and Development* (McGraw-Hill International)
[4] Ulmeanu M E and Doicin C V 2018 Development of additive manufactured products – applications of technical functional analysis (Bucharest: BREN)
[5] Doicin C V and Ulmeanu M E 2018 *Product design and development: project guide and applications* (Bucharest: BREN)
[6] Peng X, Kong L, Fuh JYH and Wang H 2021 *Journal of Manufacturing and Materials Processing* **5** 38
[7] Behera D et al. 2021 Current challenges and potential directions towards precision microscale additive manufacturing – Part IV: Future perspectives *Precision Engineering* **68** 197–205
[8] Nazir A and Jeng J-Y 2019 A high-speed additive manufacturing approach for achieving high printing speed and accuracy *Proc Inst Mech Eng C J Mech Eng Sci* **234**
[9] Lee P, Chung H, Lee SW, Yoo J and Ko J. 2014 Review: Dimensional Accuracy in Additive Manufacturing Processes *ASME: Materials; Micro and Nano Technologies; Properties, Applications and Systems; Sustainable Manufacturing* **1**
[10] Khan M A and Jappes JTW 2022 *Innovations in Additive Manufacturing* (Springer Cham)
[11] Braian M, Jimbo R and Wennerberg A 2016 Production tolerance of additive manufactured polymeric objects for clinical applications, *Dental Materials* **32** 853–861
[12] Shi Y, Yan C, Zhou Y, Wu J, Wang Y, Yu S and Ying C 2021 *Materials for Additive Manufacturing* (Elsevier)
[13] Lieneke T, Denzera V, Adam GAO and Zimme D 2016 Dimensional tolerances for additive manufacturing: Experimental investigation for Fused Deposition Modeling *Procedia CIRP* **43** 286 – 291
[14] Rupal B S, Anwer N, Secanell M and Qureshi A J 2020 Geometric Tolerance Characterization of Laser Powder Bed Fusion Processes Based on Skin Model Shapes *CIRP CAT 2020* **92** 169–174
[15] Piedra-Cascón W, Krishnamurthy V R, Att W and Revilla-León M 2021 3D printing parameters, supporting structures, slicing, and post-processing procedures of vat-polymerization additive manufacturing technologies: A narrative review *Journal of Dentistry* **109** 103630
[16] Ionescu N, Visan A, Manolache D and Nistor C 2016 *Tolerances Design* (Bucharest: PRINTECH)
[17] *** ASME Y14.5-2018 Dimensioning and Tolerancing, Engineering Product Definition and Related Documentation Practices
[18] *** ISO 1101:2017(en) Geometrical product specifications (GPS) — Geometrical tolerancing — Tolerances of form, orientation, location and run-out
[19] Ion A V 2018 Scientific coordinators Ulmeanu M E, Doicin C V, Bachelors Diploma Project, Industrial Engineering, Faculty of Industrial Engineering and Robotics, University POLITEHNICA of Bucharest, Romania
[20] Ghionea I, Visan A, Spanu P and Ionescu C 2017 *Business Inform Management Assoc* **I-IX** 966–973
[21] Ionescu C, Spanu P and Visan A 2016 *Int Business Informat Management Assoc* **I – VII** 979–992

IOP Conf. Series: Materials Science and Engineering 1268 (2022) 012005 doi:10.1088/1757-899X/1268/1/012005

Simulation of the hydraulic steering device, for a nose landing gear

B. A. Nicolin[1] and I. Nicolin[1]

[1]INCAS – National Institute for Aerospace Research "Elie Carafoli", B-dul Iuliu Maniu 220, Bucharest 061126, Romania

nicolin.adrian@incas.ro

Abstract. The hydraulic steering simulation is done with SIMULINK, part of the MathWorks MATLAB® application. All parts, subassemblies, and assemblies that define the nose landing gear (NLG) and nose wheel steering are fully defined in 3D, with CATIA V5 - a computer-aided design software used for modeling, it is possible to carry out simulations that allow preliminary evaluation, theoretically and experimentally. The purpose of this simulation is to confirm the correctness of the steering gear kinematics, that the two hydraulic cylinders have been sized correctly and can overcome the resistive steering moment, and that the steering time complies with design specifications for the nose landing gear of a military training aircraft.

1. Introduction

The work methodology for simulation using SIMULINK, part of the MathWorks MATLAB® and Simscape™ is explained in detail in various bibliographic sources [1, 20, 21, 22, 23]. The current state of research in the field of simulation applying SIMULINK, MATLAB®, and Simscape™ is widespread on a vast scale in the most varied areas of research, all over the world [2 – 17].

Simulink is a graphical editor, integrated with MATLAB®, and contains libraries of blocks that can be adapted for multidomain simulation, as well as allowing the modeling and simulation of any dynamic system. The application allows embedding algorithms from MATLAB into simulation models, exporting simulation results to MATLAB for all further analysis [20].

The main key features of Simulink® are the next [21]:

- Graphical editor for building and managing hierarchical block diagrams;
- Libraries of predefined blocks for modeling continuous-time and discrete-time systems;
- Simulation engine with fixed-step and variable-step ODE solvers;
- Scopes and data displays for viewing simulation results;
- Project and data management tools for managing model files and data;
- Model analysis tools for refining model architecture and increasing simulation speed;
- MATLAB Function block for importing MATLAB algorithms into models;
- Legacy Code Tool for importing C and C++ code into models [21].

Simscape™ enables the rapid creation of models of multidomain physical systems in the Simulink® environment. Simscape builds models of physical components based on physical connections that integrate directly with block diagrams and other modeling patterns or models. One can model varied systems from almost any domain by assembling the fundamental components into a schematic. Additional Simscape products provide more complex analysis components and capabilities [22].

Simscape allows the development of control systems and for testing system-level performance. Custom component models can be created using the MATLAB®-based Simscape language, which enables text-based creation of physical modeling components, domains, and libraries. Various models can be parameterized using MATLAB variables and expressions, and control systems can be designed for any physical system in Simulink [22].

Simulink is a powerful tool for simulation in the domain of Research and Development, and the advantages of Simulink are listed below [23]:

- Simulink is similar to hardware implementation;
- Simulink is very easy to read because it is like a working block diagram;
- The simulation is done in real-time, so the state of the entire system is available for every desired time interval [23].

2. Preparing 3D models for import into SIMSCAPE

The Nose Landing Gear (NLG) assembly is presented in figure 1. All 3D models in CATIA V5 contain a large volume of information that does not allow direct import into the SIMSCAPE module. Therefore, each subassembly in CATIA V5 will be saved as a single part using the "Generate CATPart from Product" function. Then each subassembly created with the "Generate CATPart from Product" function will be exported from CATIA V5 as a step file (name.stp), which can be imported with the SIMSCAPE module in the MathWorks MATLAB application [19].

The following functional subassemblies were created:

- a fixed subassembly, because it does not steer, but is a support for the other subassemblies;
- a rotating subassembly, which performs the steering.

Follows a group of four subassemblies that together constitute the actuator sub-assembly, i.e., the two hydraulic actuation cylinders, controlled by the MOOG 30 Series Servo Valve with the electrical connection cable over the RETURN port [18], which receives an electrical signal from the rudder pedals and the hydraulic block, according to the diagram in figure 2.

The actuator sub-assembly is divided into four subassemblies that allow the simulation of the roto-translational movement performed by the two hydraulic actuation cylinders:

- right hydraulic cylinder body subassembly;
- right piston rod subassembly with Ball joint Uniball end;
- left hydraulic cylinder body subassembly;
- left piston rod subassembly with Ball joint Uniball end.

The fixed subassembly does not perform the steering movement, but it is a support for the other subassemblies and is shown in figure 3. The fixed subassembly is exported from CATIA V5 with the name “Fixed Subassembly.stp”.

The rotating subassembly performs the steering by ±45o about the common vertical axis of the nose landing gear body, and it is shown in figure 4. The rotating subassembly is exported from CATIA V5 with the name “Rotating Subassembly.stp”.

The right hydraulic cylinder body subassembly attaches to the right-hand axle of the hydraulic block and provides support for the roto-translational movement of the piston rod with the Ball joint Uniball end, and it is shown in figure 5. The right hydraulic cylinder body subassembly is exported from CATIA V5 with the name “Right Body.stp”.

The right piston rod with Ball joint Uniball end subassembly is inserted into the right hydraulic cylinder and performs a rotational movement around the axis of rotation of the hydraulic cylinder and a translational movement along the right hydraulic cylinder axis, as is shown in figure 6. The right piston rod with Ball joint Uniball end subassembly is exported from CATIA V5 with the name “Right Rod.stp”.

The left hydraulic cylinder body subassembly attaches to the left-hand axle of the hydraulic block and provides support for the roto-translational movement of the piston rod with the Ball joint Uniball end, and it is shown in figure 7. The left hydraulic cylinder body subassembly is exported from CATIA V5 with the name “Left Body.stp”.

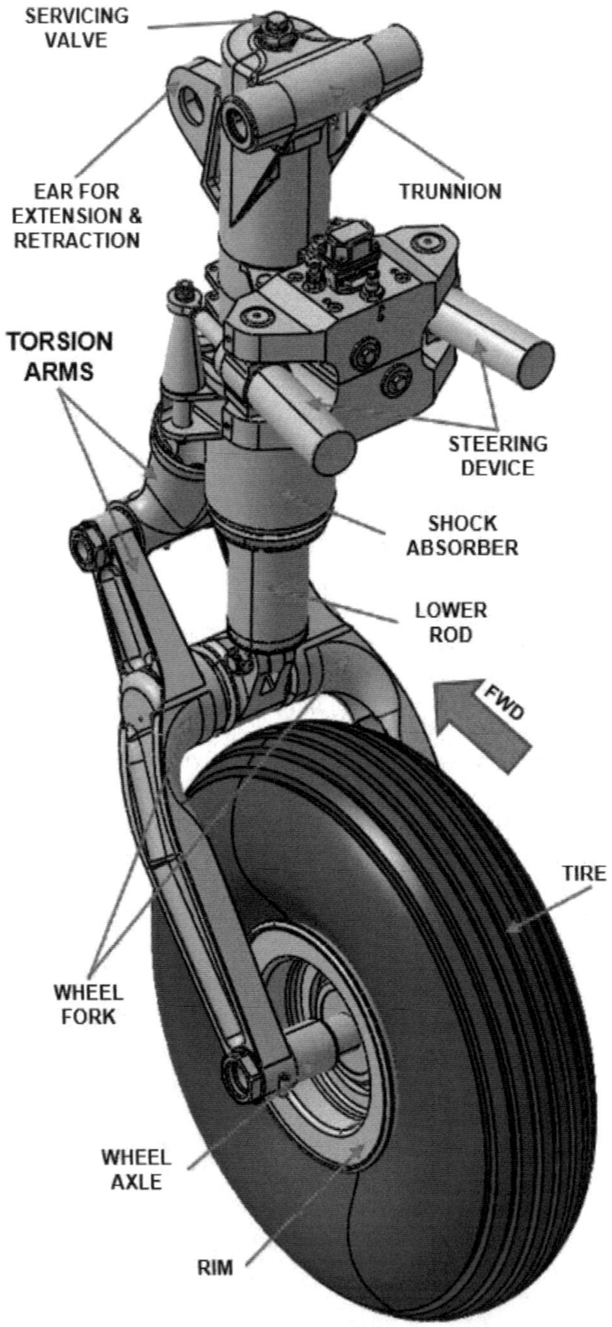

Figure 1 Nose Landing Gear (NLG) assembly.

The left piston rod with Ball joint Uniball end subassembly is inserted into the left hydraulic cylinder and performs a rotational movement around the axis of rotation of the hydraulic cylinder and a translational movement along the left hydraulic cylinder axis, which is shown in figure 8. The left piston rod with Ball joint Uniball end subassembly is exported from CATIA V5 with the name "Left Rod.stp".

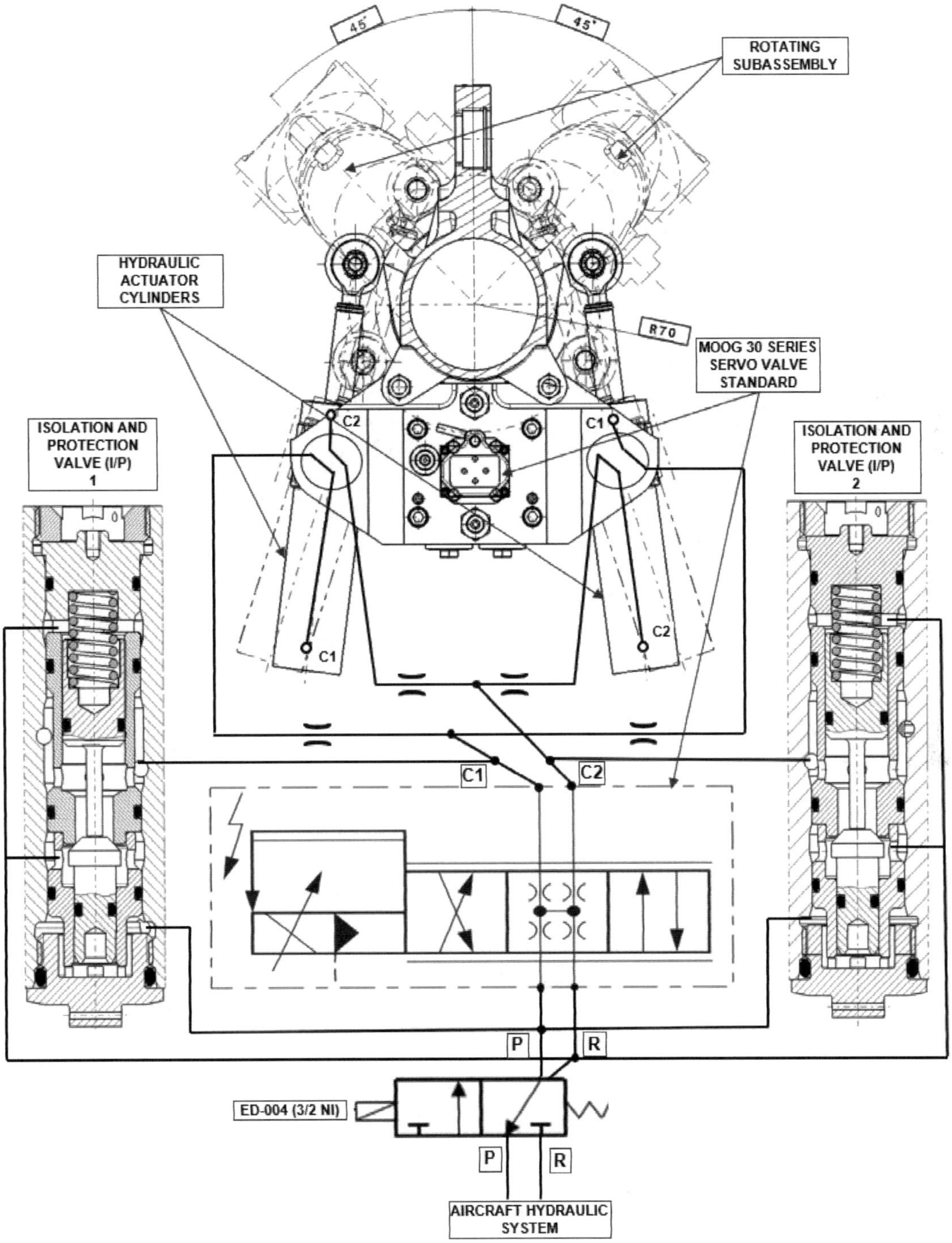

Figure 2 Concept diagram of the hydraulic circuit for steering.

IOP Conf. Series: Materials Science and Engineering 1268 (2022) 012005 doi:10.1088/1757-899X/1268/1/012005

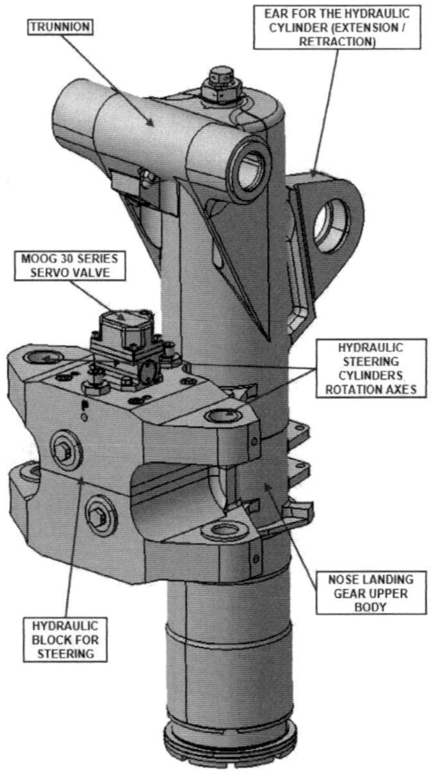

Figure 3 The fixed subassembly of the nose landing gear.

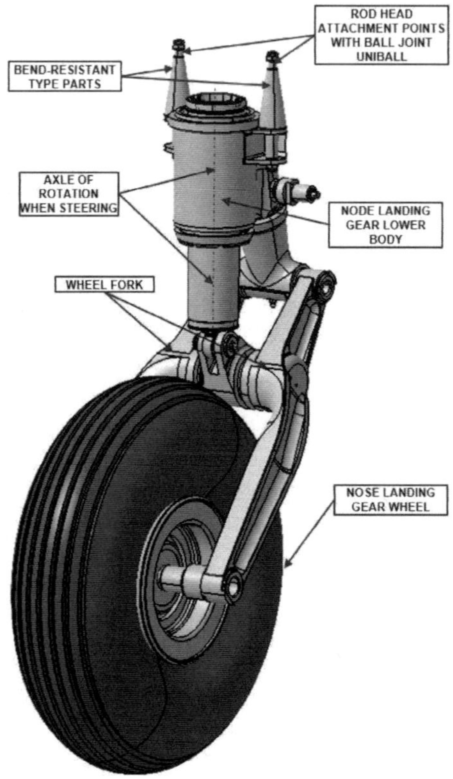

Figure 4 The rotating subassembly of the nose landing gear.

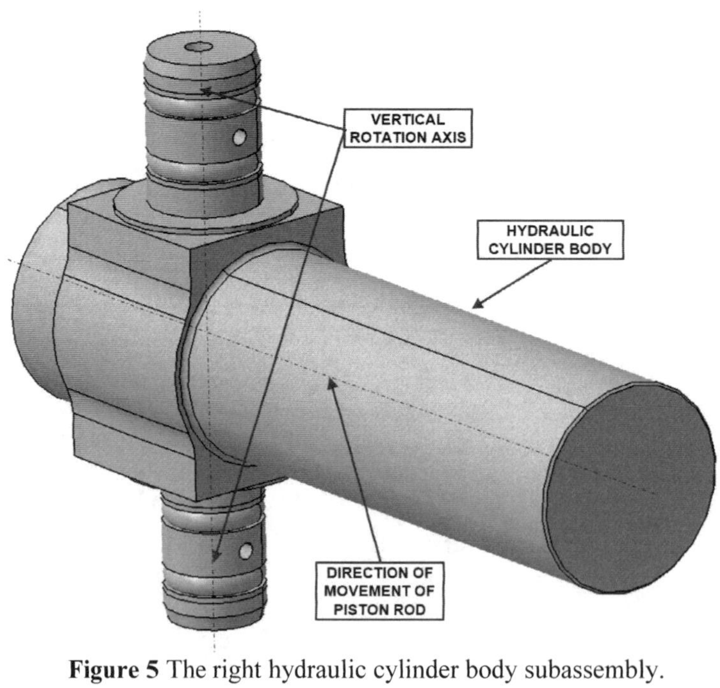

Figure 5 The right hydraulic cylinder body subassembly.

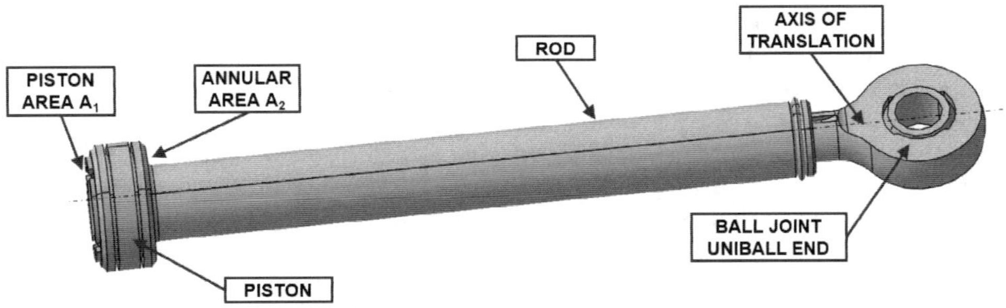

Figure 6 The right piston rod with Ball joint Uniball end subassembly.

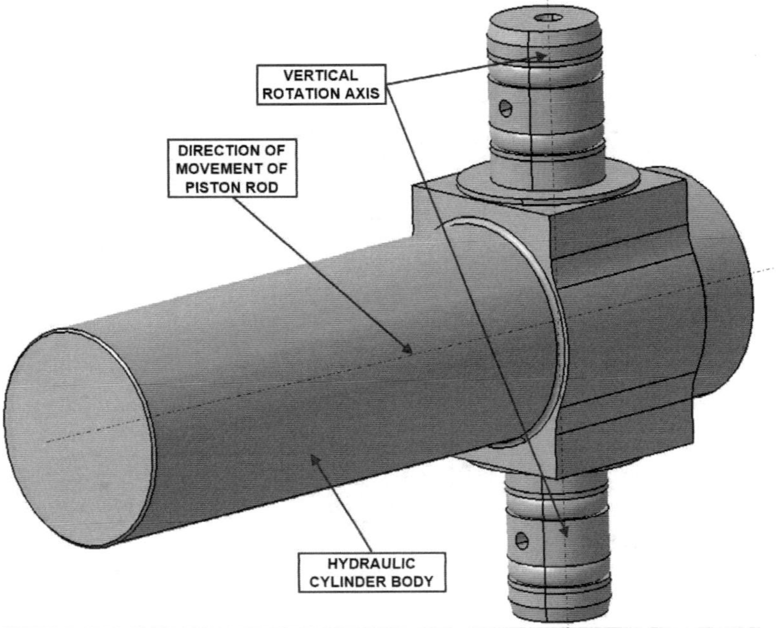

Figure 7 The left hydraulic cylinder body subassembly.

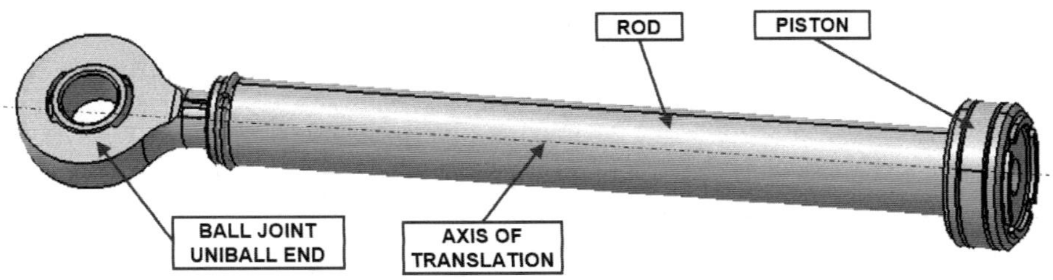

Figure 8 The left piston rod with Ball joint Uniball end subassembly.

All previously created *.stp subassemblies are imported into SIMSCAPE. The fixed subassembly is associated with the link block to the aircraft structure and the block with the description of the mechanism for which the simulation is made. Blocks are inserted between the subassemblies that describe the movement parameters and the kinematic couplings connecting them:

- rotational coupling between the fixed body and the rotating body (the one performing the turning movement);
- rotation coupling between the fixed body and each of the two bodies of the hydraulic cylinders (left and right);

- rotational coupling between the fixed body and each of the two rod-piston bodies of the hydraulic cylinders (left and right);
- translational coupling between each hydraulic cylinder body and each rod-piston body (left and right).

Blocks are added to the rotational coupling between the fixed body and the rotating body (the one performing the turning movement) with the parameters of movement, changing the direction of the steering (left or right), converting the angular movement from radians to degrees and displaying angular values during the simulation, as shown in figure 9.

The diagram in figure 1 is used to model the hydraulic supply, command, and steering control circuit. From the presented hydraulic scheme, the isolation and protection valves were not modeled because they do not play any role in simulating the steering in SIMULINK, except for the limitation of the maximum hydraulic pressure in the hydraulic cylinders of the servomechanism, to the value $p_{max} = 24.5_0^{+0.5}$ MPa. The simulation was performed at the nominal pressure of 20.7 MPa from the hydraulic system of the IAR 99NG aircraft.

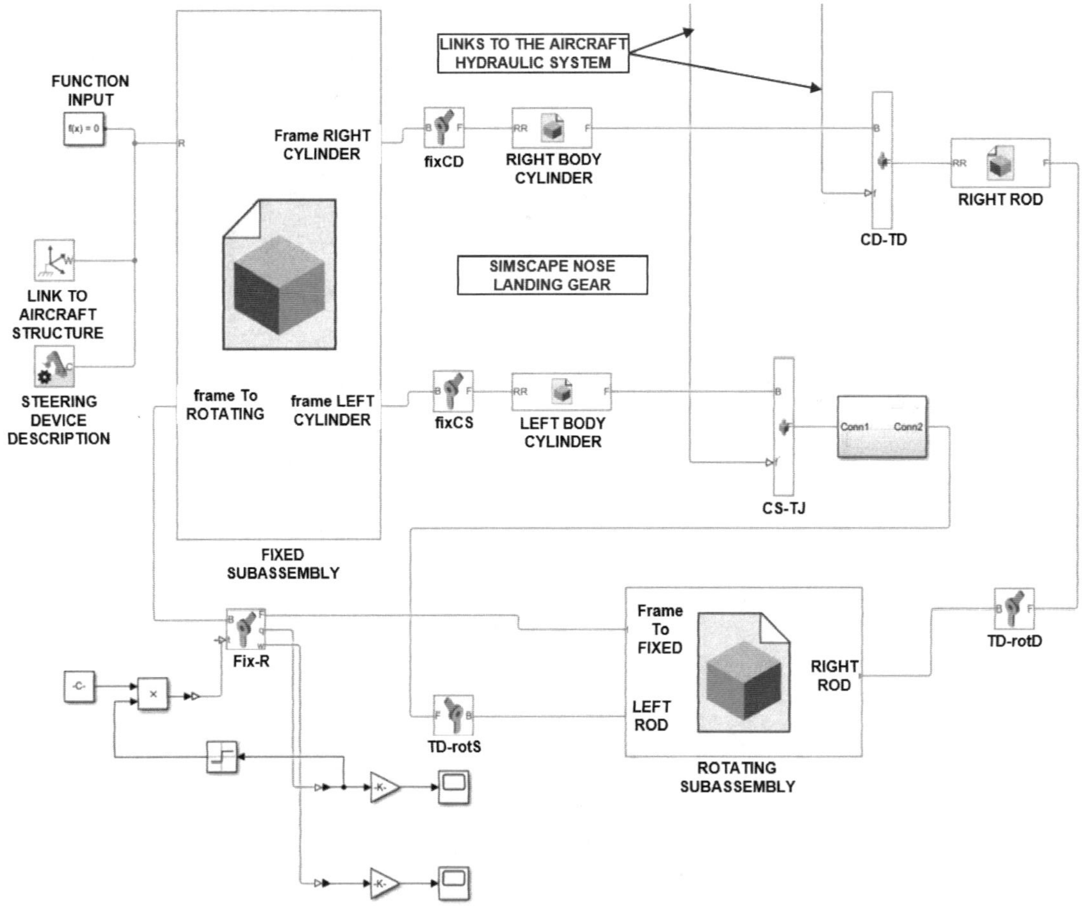

Figure 9 Scheme of connection and conditioning of subassemblies in SIMSCAPE.

The diagram of the hydraulic circuit for supply, command, and control of the steering built from existing blocks in the SIMSCAPE library is presented in figure 10. The hydraulic supply, command, and steering control circuit in SIMSCAPE includes the hydraulic fluid properties block for H-515 according to MIL-PRF-5606 [25].

The hydraulic pump for the hydraulic system of the IAR 99NG school and training aircraft is a variable flow hydraulic pump driven by the aircraft engine with a maximum flow of 36 liters/minute.

The hydraulic fluid reservoir is self-pressurized and prevents cavitation in the hydraulic pump.

The 4-way, 3-position distributor included in the standard MOOG 30 Series electrohydraulic servo valve is represented by the corresponding block to which the servo valve properties have been associated.

The two hydraulic cylinders are represented by the corresponding blocks to which specific properties have been associated: the maximum stroke is 100.4 mm (48.3 for the EXTEND stroke and 51.57 mm for the RETRACT stroke, the large piston area $A_1 = 490.625$ mm^2, and the piston rod annular area $A_2 = 314$ mm^2. Each hydraulic cylinder was attached to the rod end with a load simulator and a displacement transducer for measuring rod displacement and displacement display blocks.

Pressure transducers and pressure display blocks are located on the hydraulic pipes.

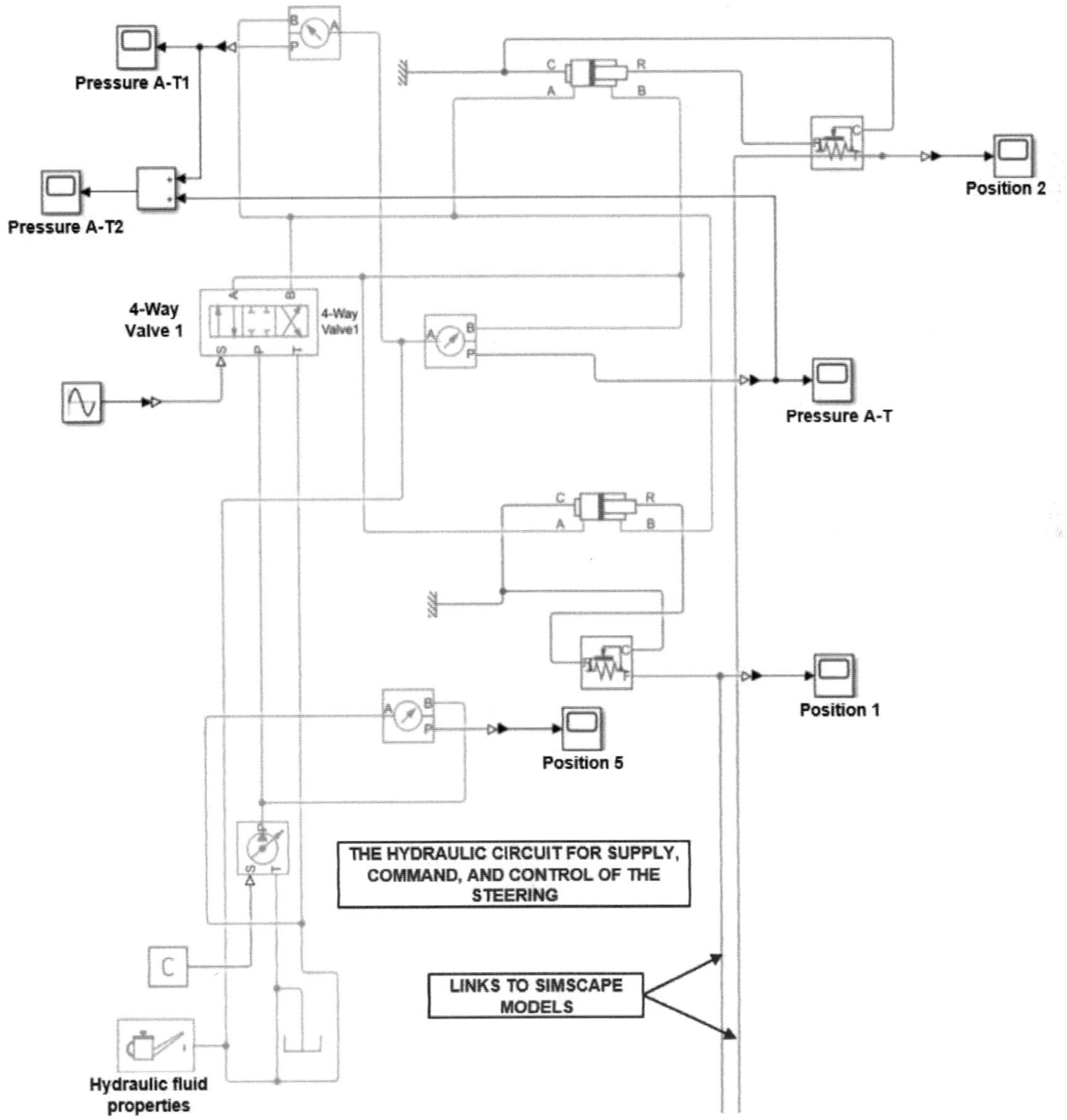

Figure 10 Scheme of connection and conditioning of subassemblies in SIMSCAPE.

Figure 11 shows the complete SIMSCAPE block diagram for simulating the entire nose landing gear steering system.

IOP Conf. Series: Materials Science and Engineering 1268 (2022) 012005 doi:10.1088/1757-899X/1268/1/012005

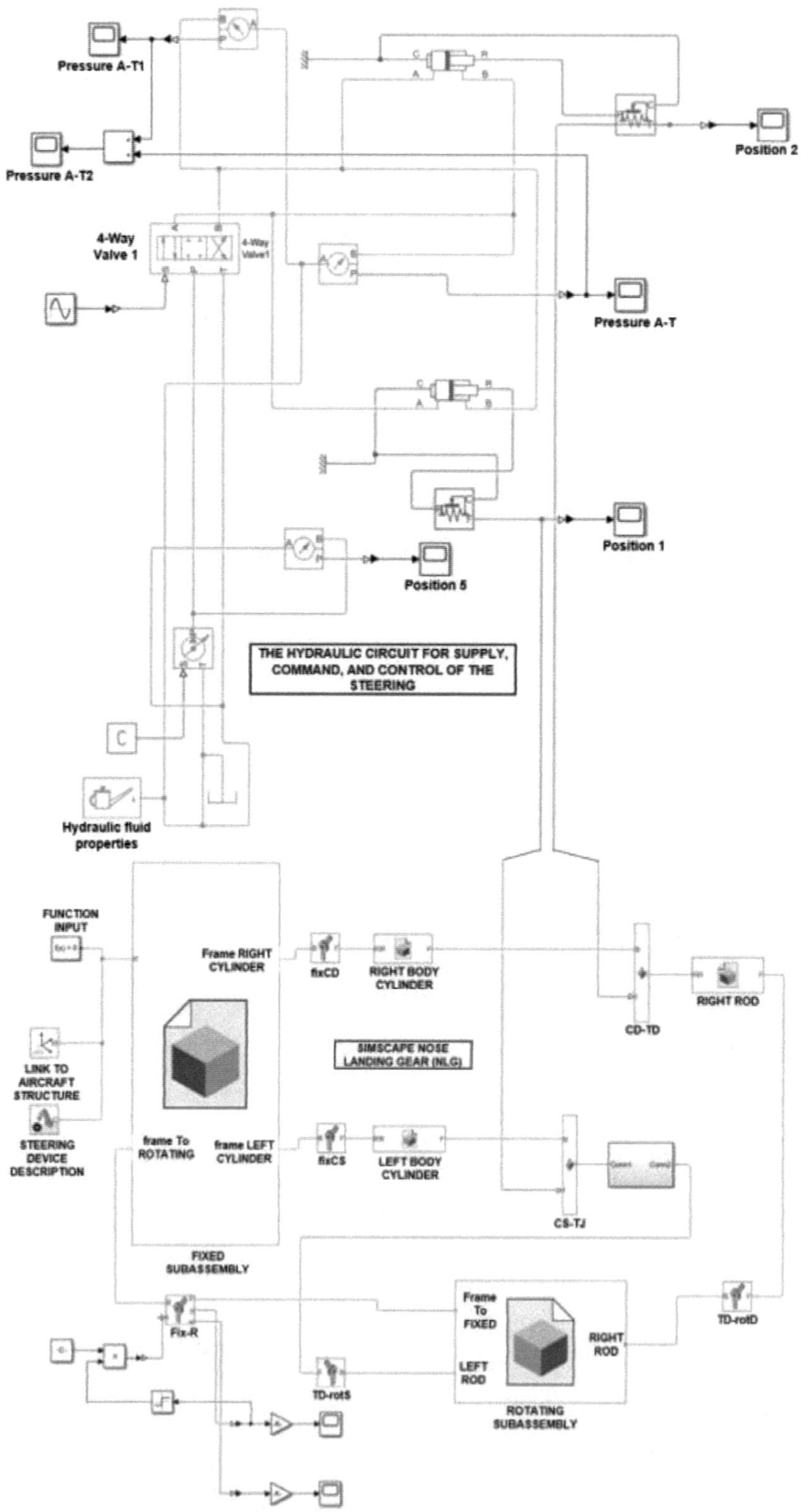

Figure 11 SIMSCAPE block diagram for simulating the entire steering system.

3. Simulation results

The simulation result is presented in a video file that recorded the simulation from SIMULINK (MATLAB): named "NLG STEERING IAR99NG.mpeg" which is attached to this document.

NLG STEERING IAR99NG.mpeg

Other results from the simulation report are presented in Tables 1 to 3.

Table 1 The resisting steering moment simulation block

Unit	Value	Lock Scale	Sample Time	Frame Period
Nm	-393,15	off	inf	inf

Table 2 Hydraulic cylinders simulation block

Piston area A_1 (Figure 6)	Area A_1 unit	Annular area A_2 (Figure 6)	Area A_2 unit	Stroke	Stroke unit	Init pos	Init pos unit	Or	Or unit
490.625	mm^2	314	mm^2	48.83+51.57	mm	51.57	mm	1	1
490.625	mm^2	314	mm^2	48.83+51.57	mm	51.57	mm	1	1

Table 3 Full list of simulation blocks

BlockType	Count	Block Names
Scope	8	Position1, Position2, Position5, Scope, Scope1, pressure A-T, pressure A-T1, pressure A-T2
PS-Simulink Converter (m)	7	PS-Simulink Converter, PS-Simulink Converter1, PS-Simulink Converter16, PS-Simulink Converter17, PS-Simulink Converter18, PS-Simulink Converter4, PS-Simulink Converter5
File Solid (m)	6	FixedAssembly , RotatingAssembly, RightCylinderBody, LeftCylinderBody, RightRod, LeftRod
Revolute Joint (m)	5	TD-rotD, TS-rotS, fix-R, fixCD, fixCS
Hydraulic Pressure Sensor (m)	3	Hydraulic Pressure Sensor, Hydraulic Pressure Sensor1, Hydraulic Pressure Sensor2
Simulink-PS Converter (m)	2	Simulink-PS Converter2, Simulink-PS Converter4
Prismatic Joint (m)	2	CD-TD, CS-TJ
Mechanical Translational Reference (m)	2	Mechanical Translational Reference1, Mechanical Translational Reference3
Ideal Force Sensor (m)	2	Ideal Force Sensor, Ideal Force Sensor1
Gain	2	Gain, Gain1
Double-Acting Hydraulic Cylinder	2	Double-Acting Hydraulic Cylinder (Simple)2, Double-Acting Hydraulic Cylinder (Simple)3

BlockType	Count	Block Names
(Simple) (m)		
World Frame (m)	1	World Frame
Sum	1	Add
Solver Configuration (m)	1	Solver Configuration
Sin	1	Sine Wave
Signum	1	Sign
Product	1	Product
PS Constant (m)	1	Pressure2
Mechanism Configuration (m)	1	Mechanism Configuration
Hydraulic Reference (m)	1	HRef1
Hydraulic Pressure Source (m)	1	Hydraulic Pressure Source1
Hydraulic Fluid (m)	1	Hydraulic Pressure Source1
Constant	1	Constant1
4-Way Directional Valve (m)	1	4-Way Valve1

4. Design specifications for the steering of the nose landing gear

The steering function for the wheel of the nose landing gear is electronically controlled to meet the design requirements as presented in figure 12.

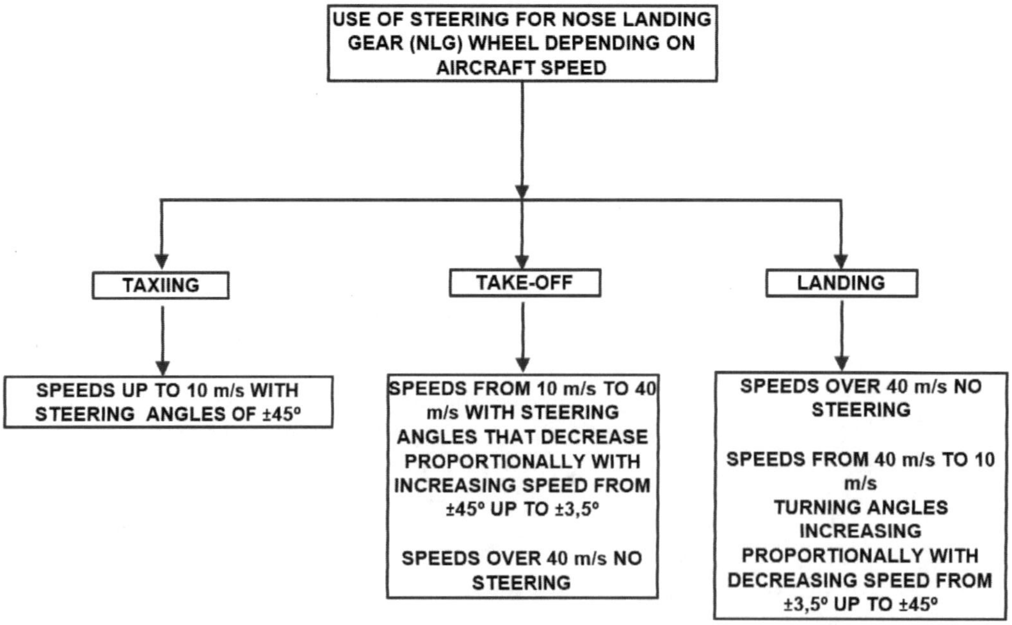

Figure 12 Use of steering for nose landing gear wheel depending on aircraft speed.

The specified angular velocity is about 90 degrees/second or $\pi/2$ radians/second. The resistant steering moment Mrs calculated (friction between the nose landing gear wheel and the runway) and imposed on the simulation is Mrs = 393,15 [Nm].

5. Conclusions

The simulation results obtained up to this point confirm:

- the kinematics of the steering device were well calculated because the simulation of the steering device works very well;
- the two hydraulic cylinders were correctly dimensioned and can overcome the resistant steering moment even if it is double the calculated resistant moment;
- the steering time complies with the specified requirements for the steering of the nose landing gear of a military training aircraft.

Future research directions will include the analysis of the functional parameters that were recorded by the measuring devices included in the SIMSCAPE block diagram, such as:

- the variation of the maximum active torque MD(γ) to perform the steering depending on the steering angle $\gamma = 0 \ldots 45^{o}$;
- the required hydraulic pressure depending on the steering angle $\gamma = 0 \ldots 45^{o}$;
- hydraulic fluid temperature depending on the actuation time.

Acknowledgment
The work was carried out within the project NUCLEU, contract no. 8N/07.02.2019, Additional Act no. 11/2022, supported by the Romanian Ministry of Research, Innovation, and Digitization.

References

[1] Xue, Dingyü. *Modeling and Simulation with Simulink®: For Engineering and Information Systems*, Berlin, Boston: De Gruyter, 2022. https://doi.org/10.1515/9783110734959.
[2] Baloi, A., et al. *Advantages of Using MatLab Simulink in Laboratory Lessons on Operating Conditions of Overhead Power Lines*. Procedia - Social and Behavioral Sciences, Volume 191, 2015, Pages 179-184, ISSN 1877-0428, https://doi.org/10.1016/j.sbspro.2015.04.367.
[3] Ashok Kumar L., et al. *Chapter 1 - MATLAB®/Simulink, Software Tools for the Simulation of Electrical Systems*, Academic Press, 2020, Pages 1-35, ISBN 9780128194164, https://doi.org/10.1016/B978-0-12-819416-4.00001-6.
[4] Omer Eldirdiry., et al. *Modeling of a biped robot for investigating foot drop using MATLAB/Simulink,* Simulation Modelling Practice and Theory, Volume 98, 2020, 101972, ISSN 1569-190X, https://doi.org/10.1016/j.simpat.2019.101972.
[5] Falcone, A., et al. *A Simulink-based software solution using the Infinity Computer methodology for higher order differentiation*, Applied Mathematics, and Computation, Volume 409, 2021, 125606, ISSN 0096-3003, https://doi.org/10.1016/j.amc.2020.125606.
[6] Basana, F., Branz, F. *Simulation of robotic space operations with minimum base reaction manipulator*, Journal of Space Safety Engineering, Volume 9, Issue 3, 2022, Pages 440-448, ISSN 2468-8967, https://doi.org/10.1016/j.jsse.2022.06.005.
[7] Žáková, K., Balogh, R. Control Engineering Home Learning Kit, IFAC-papers online, Volume 55, Issue 4, 2022, Pages 310-315, ISSN 2405-8963, https://doi.org/10.1016/j.ifacol.2022.06.051.
[8] Bârsan, A., et al. *Dynamic analysis of a robot-based incremental sheet forming using Matlab-Simulink Simscape™ environment*, Materials Today: Proceedings, Volume 62, Part 5, 2022, Pages 2538-2542, ISSN 2214-7853, https://doi.org/10.1016/j.matpr.2022.03.134.
[9] Falcone, A. et al. *A Simulink-based software solution using the Infinity Computer methodology*

for higher order differentiation, Applied Mathematics, and Computation, Volume 409, 2021, 125606, ISSN 0096-3003, https://doi.org/10.1016/j.amc.2020.125606.

[10] Bacharoudis, K. et al. *An efficient cost estimation framework for aerospace applications using Matlab/Simulink*, Procedia CIRP, Volume 104, 2021, Pages 1143-1148, ISSN 2212-8271, https://doi.org/10.1016/j.procir.2021.11.192.

[11] Salazar-Herran, E., et al. *Flexible dynamic model of PHEX for transient simulations in Matlab/Simulink using finite control volume method*, International Journal of Refrigeration, Volume 110, 2020, Pages 83-94, ISSN 0140-7007, https://doi.org/10.1016/j.ijrefrig.2019.11.003.

[12] Piedrafita, R., et al. *Simulink® implementation and industrial test of Input Shaping techniques*, Control Engineering Practice, Volume 79, 2018, Pages 1-21, ISSN 0967-0661, https://doi.org/10.1016/j.conengprac.2018.06.021.

[13] Martyanov, A.S., et al. *Development of Control Algorithms in Matlab/Simulink*, Procedia Engineering, Volume 129, 2015, Pages 922-926, ISSN 1877-7058, https://doi.org/10.1016/j.proeng.2015.12.135.

[14] Palkovič, L., et al. *Integration of Inertial Measuring Unit Platform into MATLAB Simulink*, IFAC Proceedings Volumes, Volume 45, Issue 11, 2012, Pages 200-205, ISSN 1474-6670, ISBN 9783902823014, https://doi.org/10.3182/20120619-3-RU-2024.00089.

[15] Sivák, P., Hroncová, D. *State-Space Model of a Mechanical System in MATLAB/Simulink*, Procedia Engineering, Volume 48, 2012, Pages 629-635, ISSN 1877-7058, https://doi.org/10.1016/j.proeng.2012.09.563.

[16] Radionova, L.V., Chernyshev, A.D. *Mathematical Model of the Vehicle in MATLAB Simulink*, Procedia Engineering, Volume 129, 2015, Pages 825-831, ISSN 1877-7058, https://doi.org/10.1016/j.proeng.2015.12.114.

[17] Dejan Ž. Jokić, Slobodan D. Lubura, Duško Lukač, *Development of integral environment in Matlab/Simulink for FPGA*, IFAC Proceedings Volumes, Volume 46, Issue 28, 2013, Pages 50-55, ISSN 1474-6670, ISBN 9783902823533, https://doi.org/10.3182/20130925-3-CZ-3023.00003.

[18] Nicolin, B. A., Scientific report no. 5, "Politehnica" University Bucharest, Doctoral School of Industrial Engineering and Robotics, 2022.

[19] Nicolin, I., Nicolin, B. A., et al. Technological design for a new generation of school and training aircraft IAR-99NG. Project NUCLEU, contract no. 8N/07.02.2019, Additional Act no. 11/2022. INCAS Bucharest, Romania, 2019-2022.

[20] * * * https://www.mathworks.com/help/simulink/index.html?s_tid=CRUX_lftnav

[21] * * * https://www.mathworks.com/help/simulink/gs/product-description.html

[22] * * * https://www.mathworks.com/help/simscape/?s_tid=srchbrcm

[23] * * * https://www.mathworks.com/products/simulink.html

[24] * * * https://www.moog.com/products/servovalves-servo-proportional-valves/industrial/flow-control/analog-without-integrated-electronics/30-series-flow-control-servo-valve.html

[25] MIL-PRF-5606H. Hydraulic Fluid, Petroleum Base; Aircraft, Missile, and Ordnance.

IOP Conf. Series: Materials Science and Engineering 1268 (2022) 012006 doi:10.1088/1757-899X/1268/1/012006

Properties of photo-curable polyurethane resins used in SLA manufacturing

D.I. Băilă[1]*, S. Tonoiu[1]

[1] University Politehnica of Bucharest, 313 Splaiul Independentei, Bucharest, 060042, Romania

Corresponding author: baila_d@yahoo.com

Abstract. Stereolithography (SLA) is the oldest additive manufacturing technology, and it is a 3D printing technology used to create industrial concept models, rapid prototypes or different parts with complex geometries in a very short time. By SLA technology, it is possible to manufacture parts with extremely high feature resolutions and quality surface finishes, using a wide selection of materials. SLA 3D printing is frequently used for manufacture: the parts requiring high accuracy and features as small as 0.07 mm, smooth surfaces with good rugosities and form and fit testing. In this article were presented the mechanical properties of the photo-curable polyurethane resins used in SLA technology, the SEM analysis and for establish the composition contain, it was used the ATR-FTIR analysis. In conclusion, new photo-curable polyurethane resins, used in SLA technology, allow the creation of parts with a high degree of accuracy, with good mechanical properties, which can be used both in industry and in medicine domain.

1. Introduction

Stereolithography (SLA or SL) is an additive manufacturing technology known as a manufacturing process by solidification by photopolymerization, of the raw material in the state liquidate. Stereolithography was the first process that allowed the generation of a physical model, using model data, directly from the computer. The frequently used technology in additive manufacturing is Stereolithography (SLA).

The stereolithography SLA technology uses a photopolymer resin that can solidify under the action of rays UV transmitted by an ultraviolet laser to build the layers of the 3D object.

The desired 3D model is initially sliced into cross sections. For each layer, the laser beam draws a cross section of the partial model on the surface of the liquid resin. Exposure to the ultraviolet laser light solidifies the model drawn on the liquid resin resulting in a solid layer constructed (3D printed) which is added to the previously constructed layer.

After the pattern has been drawn, the platform descends with a distance equal to the thickness of a single layer, typically between 0.05 mm and 0.15 mm.

Stereolithography (SLA or SL) is a widely used additive manufacturing technology in the industrial environment for making molds, models and even functional components. Also known as photo-solidification or optical fabrication, stereolithography involves using a laser beam of ultraviolet light to solidify a resin liquid photopolymer located in the construction tank of the printer. The photopolymerization process of liquid resin (sensitive to ultraviolet light) it is realized under the action of ultraviolet light laser by solidification of the successive layers thus obtaining the solid 3D model. The technology applied is quite expensive, which leads to quite high costs for type printers SLA. [1,2]

The accuracy of the printed parts is very good, the finish of the printed surfaces is very good, print speed is good to very good. Materials used are photosensitive liquid resins and ceramic materials (newly developed). SLA technological advantages are parts prototyping with complex and extremely detailed geometries, with very great accuracy of printed surfaces, large part build sizes and the printed parts can be used as a master mold for injection molding, thermoforming, metal casting and parts industries resistant to high temperatures. [1-3]
Disadvantages of SLA technology consist of average resistance to mechanical processing, durability decreased over time, long exposure to the sun damages the parts which become brittle, require troublesome post-processing operations (with potentially hazardous chemicals). [4,5]
Also, the high cost of the printer is another disadvantage. Liquid resins can be toxic, ventilation being mandatory. After the construction is completed, the obtained 3D model is immersed in a separate chemical bath to remove excess resin after which it is treated in an ultraviolet oven for final curing.
The parts are solidified in the presence of the laser, at low laser powers (5-10 W). Most SLA printers can produce parts with sizes of 50x50x60 cm, but there are also SLA – Mammoth stereolithography printers, which can make large parts large with a construction platform of 210x70x80 cm. [6-8]
The novelty of the paper consists in establishing the mechanical properties of polyurethane photopolymerizable resins, the chemical composition by performing FTIR-Raman analyzes and the 3D printing of parts used in the industrial field, which requires very fine accuracy.

2. Experimental parts

For the experimental parts, it was used a 3D printer Formlabs Form 3+.
The Formlab Form3 printer is a modern printer that produces parts with an accuracy of 25-300 microns. It is equipped with a low power laser (P=250 mW and λ=405 nm). The software used is Preform. The file types used are STL, OBJ or FORM. This 3D printer can used different resins to manufacture, as: draft resin, greyscale resin, clear resin, colour kit and grey pro resin.
Draft Resin prints up to four times faster by rapport with the standard Formlabs materials and it is recommended for the prototypes and rapid iterations to help bring products to market faster. Parts printed with Draft Resin feature a smooth gray finish and high precision. Use the 200 microns settings for fast print speeds or use the 100 microns settings for finer detail designs.
In Figure 1, it is presented the fixing clip part manufactured by SLA technology. In Figure 1.a) the STL file is introduced in Preform software to establish the 3d print parameters necessary to manufacture the parts, and the supports necessary. In the figure 1.b), it can remark the parts printed 3D by SLA technology. In figure 1.c), it presents cardiovascular stents with a high accuracy manufactured via SLA technology. In figure 2 it is shown the manufacture of fixing clip part on Formlabs 3D printer.

a) b) c)

Figure 1. The SLA process of fixing clip part: a) Preform software used; b) printed 3D part; c) cardiovascular stents manufactured via SLA

Figure 2. The manufacturing process of fixing clip part, using Formlabs 3D printer

In the Table 1, are presented two important parameters for the manufacturing process, meaning the heat deflection and elongation at break for the different types of polyurethane resins used in SLA technology, as: standard, high temp, rigid, grey pro and tough.

Table 1. The mechanical properties of the polyurethane resins used on the Formlabs 3D printers [5]

Material	Heat Deflexion Time	Elongation at break (Post-Cured)
High Temp	238°C	2.3%
Standard	73.1°C	6.2%
Rigid	88°C	5.6%
Grey Pro	77.5°C	13%
Tough	48.5°C	24%

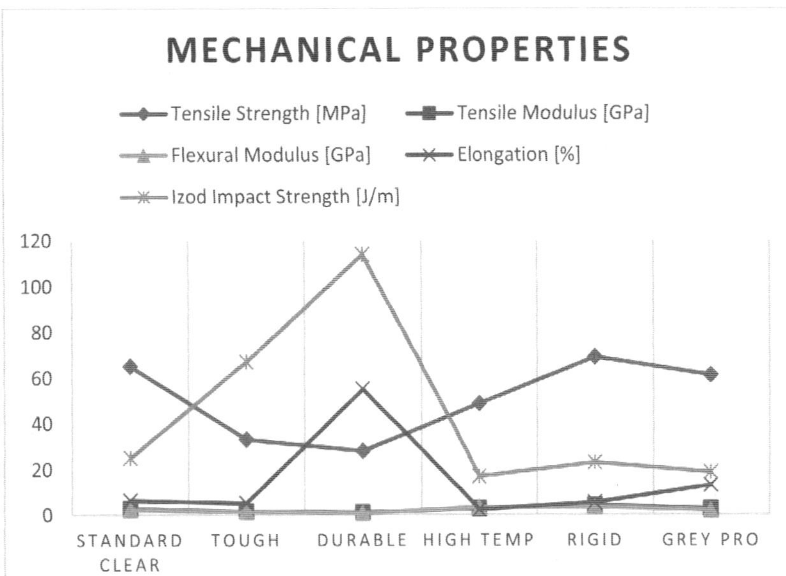

Figure 3. The mechanical properties of the different types of resins used by the Formlab Form3+ printer [5]

The Preform software was used to prepare the 3D printing of the part. The part had 1030 layers, the printing duration was 4 h and 18 min and around 20.97 ml of photopolymerizable resin was consumed. The material used was a gray photopolymerizable resin, with special mechanical resistance, used for the manufacture of prototypes and models in the field of mechanical engineering that require high rigidity. The mechanical properties of the different types of polyurethane resins used by the Formlab Form3+ printer is presented in Figure 3, like: tensile strength [MPa], tensile modulus [GPa], flexural modulus [GPa], elongation [%], IZOD impact strength [J/m] and heat deflection [°C].

3. Results and discussions

The morphology of photocurable resins was determined by a scanning electron microscope (SEM) Quanta 250, fabricated by FEI Company. The preparation of the samples was minimal and consisted in immobilizing on a double-sided carbon tape, without coating. Diffuse reflectance UV-VIS spectra were recorded using a spectrophotometer Perkin Elmer Lambda 35. The certified reflectance standard has been used and the measurements were carried out in the range 900-200 nm.

Figure 4. SEM image of surfaces of the SLA resin material

In figure 4 can remark the uniform structure and very smooth surface with fine rugosity of the photocurable resins used in SLA technology.

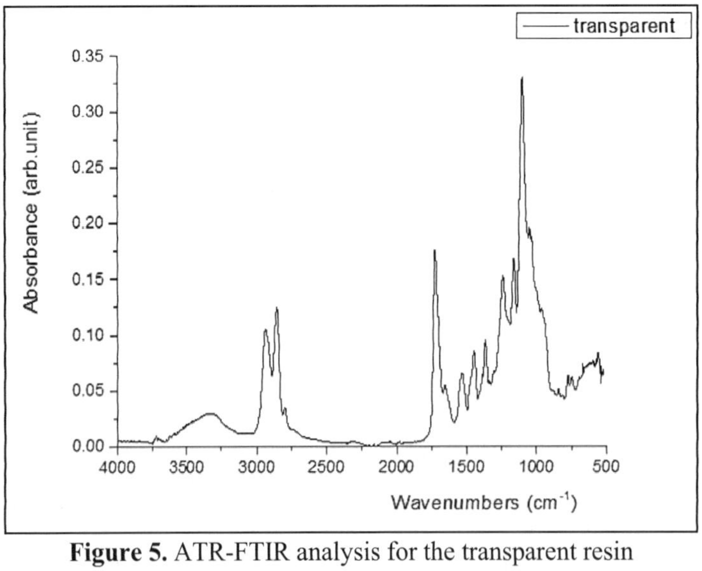

Figure 5. ATR-FTIR analysis for the transparent resin

The structure of polyurethane photo-curable resins was investigated using FT-IR Spectroscopy with a Nicolet 6700 apparatus in 400-4000 cm^{-1} domain, and with sensibility of 4 cm^{-1}. Perkin-Elmer STA

6000 apparatus was used for DTA/TG analyses, in the air atmosphere, between 20-900°C. The main bands in the infrared spectra, the antisymmetric, vas, and symmetric stretching, vs, of bonds in the molecules of polyurethane photo-curable resins are presented in the figure 5 for the clear resin (mixture between: urethane dimethacrylate 55-75%wt.; methacrylate monomer 15-25%wt.; diphenyl (2,4,6-trimethylbenzoyl) phosphine oxide <0.9%wt.), figure 6 for grey pro resin (mixture between: methacrylate monomers 30-50%wt.; urethane dimethacrylate 55-75%wt.; photoinitiators<1.5%wt.), in the figure 7 for Castable wax 40 resin – blue (mixture between: urethane dimethacrylate 35-55%wt; acrylic esters 10-20%wt; photoinitiators <2%wt.) and in figure 8 for Castable Wax resin – mauve (urethane dimethacrylate 60- 80%, additives).

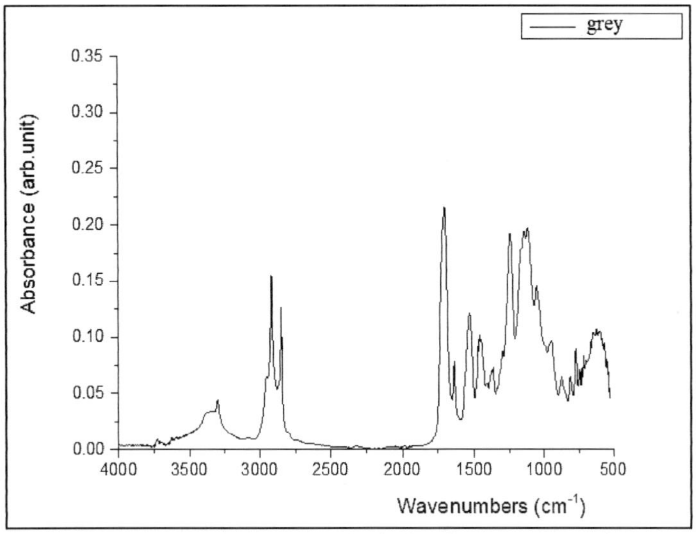

Figure 6. ATR-FTIR analysis for the gray pro resin

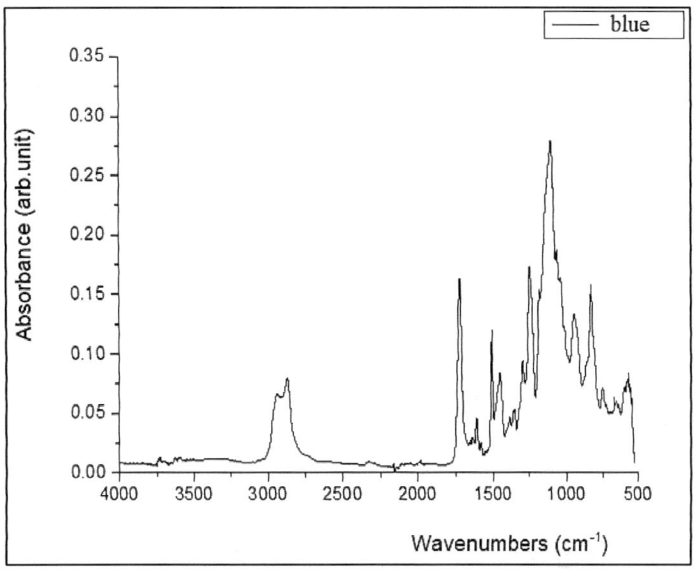

Figure 7. ATR-FTIR analysis for the blue resin

Conforming the ATR-FTIR analysis, we can remark that these resins have the same compositions, the only difference consists in additives used for color and for improve the mechanical properties, presented in figure 3. The polyurethane resins present high transmittance in infrared, and can conclude that present good optical properties.

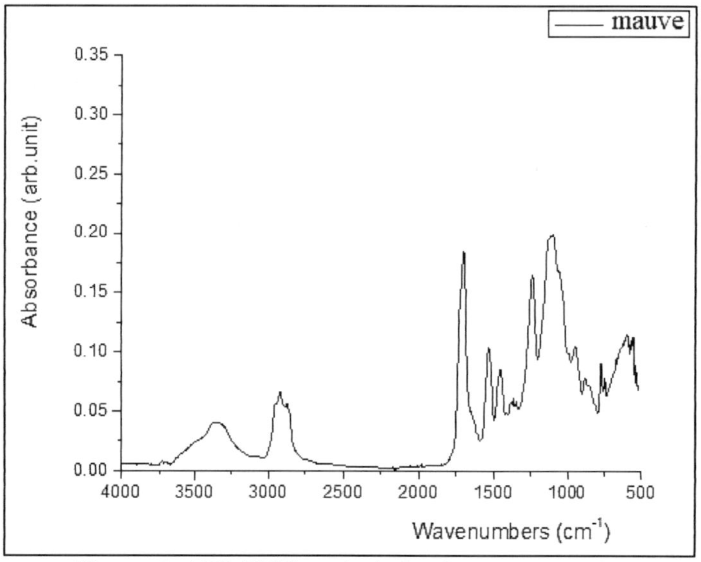

Figure 8. ATR-FTIR analysis for the magenta resin

4. Conclusions

Grey pro resin is a versatile material, presenting a high precision, moderate elongation, and resistance to deformation, and it is recommended for engineering applications. This material support print resolutions between 50 and 100 μm. Clear resin presents optical transparency and can be used in fluidics and moldmaking and in optics and lighting products. Castable Wax Resin accurately captures intricate features and offers smooth surfaces, and for this advantages, this resins it is used in the stereolithography technology for dentistry and jewelry domains. Castable Wax 40 resin has blue color, and it is used for highly detailed designs and offers surface smoothness, being used for fine detailed surfaces products. The SEM analysis shows the surface smoothness and their uniformity. In conclusion, the ATR-FTIR analysis confirms the urethane dimethacrylate presence in all photo-curable resin samples, presented in the article research, the SEM analysis realized show a smooth surface and a fine roughness, meaning that the different types of polyurethane resins used in SLA technology cand be used successfully to manufacture industrial and medical complex parts with a very fine accuracy, presenting good mechanical properties.

References

[1] Huang J, Qin Q, Wang J 2020 *Processes* **8**, 1138
[2] Cooke M, Fisher J, Dean D, Rimnac C, Mikos A 2002 *J. Biomed. Mater. Res. Part B Appl. Biomater.* **64**, 65–69.
[3] He Y, Wu Y, Fu J, Gao Q, Qiu J 2016 *Electroanalysis* **28**, 1658–1678
[4] Wallin T, Pikul J, Shepherd R 2018 *Nat. Rev. Mater.* **3**, 84–100
[5] https://formlabs.com/eu/materials/
[6] Ahn J, Huh J, Choi J 2019 *The Journal of Korean Academy of Prosthodontics* **57,** 2, 110
[7] Bartolo P, Gaspar J 2008 *CIRP Ann.* **57**, 235–238
[8] Han D, Yang C, Fang N, Lee H 2019 *Addit. Manuf.* **27**, 606–615.

IOP Conf. Series: Materials Science and Engineering 1268 (2022) 012007 doi:10.1088/1757-899X/1268/1/012007

Vehicles fleet communications in data infrastructure unavailability situations

T Savu[1] and A B Jugravu[1]

[1] Manufacturing Engineering Dept - TCM, University POLITEHNCA of Bucharest, 313 Spl. Independentei, sector 6, Bucharest, 060042, Romania

tom.savu@upb.ro

Abstract. The paper describes a solution for operating a fleet of AGVs or AMRs, when an appropriate communication infrastructure is not available, by using messages send from one vehicle to another. All vehicles are equipped with two-way radios and a server (fixed or placed on one vehicle) is managing the network messages and is storing the data about the position of each vehicle. A set of rules was defined for message management and the messages format was defined in detail. It was also defined an algorithm, running on the server, for defining the chain of vehicles to be used for sending one message in the network and for updating the data about the vehicles' network structure. It was developed and tested the software to be implemented in the radio communication modules for managing the messages flow. Tests were performed for assessing the coverage area in a real environment and the results, proving the correct software implementation, are explained.

1. Introduction

When operating a fleet of AGVs (automated guided vehicles) or AMRs (autonomous mobile robots), a key aspect is represented by the data communication infrastructure, which is used both for dispatching orders to the robots and for receiving data about their status and about the performed activities. There are numerous descriptions of vehicle-to-vehicle communication systems, with different solutions concerning their structure, communication means and performance characteristics [1]. A lot of literature exists on the topics of fleet communications and uncertain situations. Some papers are concentrating on the aspects which are influencing the communications performance [2], while others are describing various methods for planning each vehicle path under different uncertainties [3] or under changing environmentally conditions [4], for reducing the latency of the communication [5]

It is presumed all the time that communication exists between any two vehicles in the fleet [6], [7] and also the distance between any two vehicles can be measured [8]

There are cases when the vehicles fleet has to be deployed, but the infrastructure is not suitable, does not exist or is not an option to be implemented, either from technical or budget related reasons. An example of such a case would be a group of buildings (a factory or a campus) where the vehicles have to travel both inside the buildings and between them. If a Wi-Fi network exists in the complex, it will somehow solve the problem only on relatively small areas around the access points but placing access points in the areas between the buildings, over large distances, may prove to be unreliable. Even inside the buildings, there is not a simple task to maintain a network of access points which will allow the vehicles to communicate when moving from one point to another [9]. More relevant examples are those when the vehicles fleet has to be deployed over large areas, such as agricultural ones, where the GPRS signal does not exist. Last but not least, a typical example of lack of infrastructure is that when a power outage appears in an area, so any existing communication network will not function.

The solution described in this paper can be applied to the situations when, from different reasons, the vehicles fleet can rely only on its own resources to communicate, which means that messages are to be sent and received using only data communication modules installed on the fleet's vehicles.

2. Network message types

The fleet is presumed to be composed from a number of vehicles and one server. The server may be fixed, or, in more complex cases, it can reside on one of the vehicles. In the latter case, the fleet may have the ability to assign the server role to another vehicle, if the previous one is compromised.

To allow communication over large distances, both the server and the vehicles are equipped with two-way radios. All the data sent by the vehicles is stored on the server. If the distance between a vehicle and the server is larger than the radio communication range, or if different obstacles are not allowing the direct communication, data from the distant vehicle has to arrive to the server using a chain of other vehicles functioning as relays.

In such type of networks, it is essential that two or more devices are not emitting at the same time. For this, a vehicle has to answer to a message only when it is asked to, so each message in the network has to specify which is the vehicle which has to answer.

It may happen that a message is send and the vehicle to which the message is addressed is not reachable. In this situation, the network does not have to wait indefinitely for the answer, so each message will have a validity period. If the not reachable vehicle was asked by the server, the server can manage this situation directly and send the next message. If the breakage appears inside a chain of vehicles, the last vehicle which emitted will have, after a specified timeout, to send back a message mentioning the event.

Following these rules, the format of the messages is defined as [DEST] [SOURCE] [CMD] [DATA], where

- [DEST] is the ID of the destination device, server, or vehicle.
- [SOURCE] is the ID of the device sending the message.
- [CMD] is the message type code, which can be one of the following:
  - GET, when the destination device has to return its data.
  - ASK, when the destination device has to forward the message.
  - DAT, specifying that the source device is sending device data or an error code.
  - ERR, specifying that an error appeared.
- [DATA] is a field whose content depends on the message type code:
  - if the type code is GET, the DATA field is empty.
  - if the type code is ASK, the DATA field contains one or more device IDs, comma separated.
  - if the type code is DAT, the DATA field contains device data from the destination device or from another device, depending on the GET or ASK code of the message initially sent by the server.
  - if the type code is ERR, the DATA field contains an error code and the ID of the device which generated the error.

For the example from figure 1, when asking for data from vehicle V6, the server will send the message V2 SRV ASK V3,V5,V6 and the further messages sent by the vehicles will be those from table 1, where the [DATA] sections will contain the data values from V6.

Table 1. Vehicle generated messages in a transmission chain.

Sender	Message	Sender	Message
V2	V3 V2 ASK V5,V6	V5	V3 V5 DAT [DATA]
V3	V5 V3 ASK V6	V3	V2 V3 DAT [DATA]
V5	V6 V5 GET	V2	SRV V2 DAT [DATA]
V6	V5 V6 DAT [DATA]		

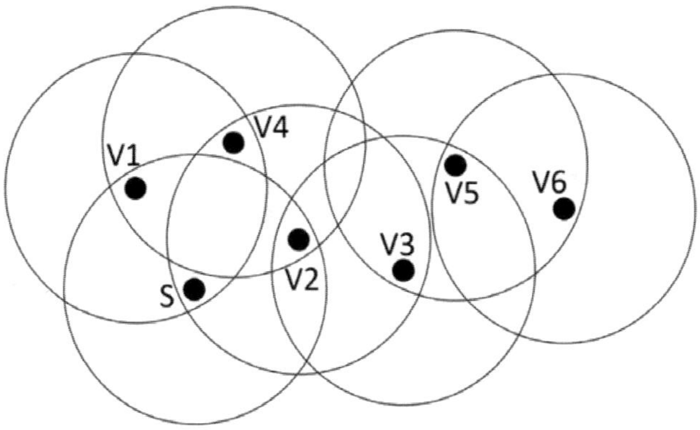

Figure 1. Network structure example (circles represent coverage areas).

If vehicle V5 will move away and will lose its connection with V3, the messages will look like in table 2, where the [DATA] sections will indicate that the V3 – V5 connection was lost.

Table 2. Transmission chain messages when a vehicle is unreachable.

Sender	Message
V2	V3 V2 ASK V5,V6
V3	V5 V3 ASK V6
V3	V2 V3 ERR [DATA]
V2	SRV V2 ERR [DATA]

3. Server's search algorithm

Initially, the server performs a round-robin data request from all vehicles, without having any information about their positions. The vehicles which will answer will be those inside the server's coverage area. These will be called directly connected vehicles. In the above-mentioned example, these will be vehicles V1 and V2, so the graph will look like in figure 2.

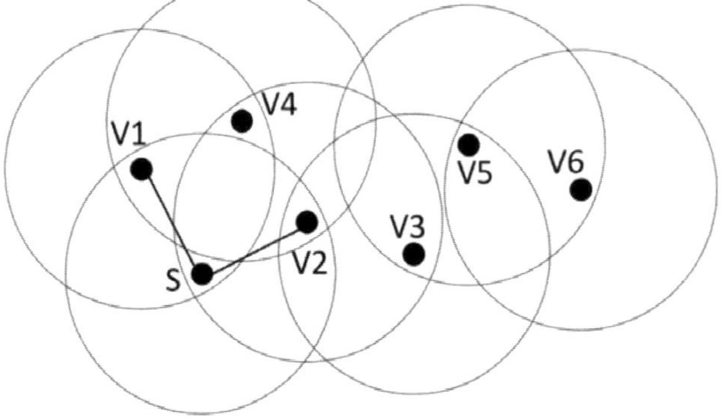

Figure 2. Graph links for directly connected vehicles.

After the initial round-robin data request, if there are vehicles still not contacted, the server will ask the directly connected vehicles to look after the missing ones. If CV1, CV2, …, CVm are the identifiers of the m directly connected vehicles and MV1, MV2, …, MVn are the identifiers of the missing vehicles, this stage of the algorithm will have the following steps:

Step 1: for (j = 1; j <= n; ++j)
Step 2: for (i = 1; i <= m; ++i)
Step 3: If CVi detects MVj (save the link; jump to Step 1) else (jump to Step 2)

In figure 3, links from the server to vehicles V1 and V2 were determined during the initial round-robin data request. Then, V2 was the first vehicle to connect to V3 and V1 was the first vehicle to connect to V4.

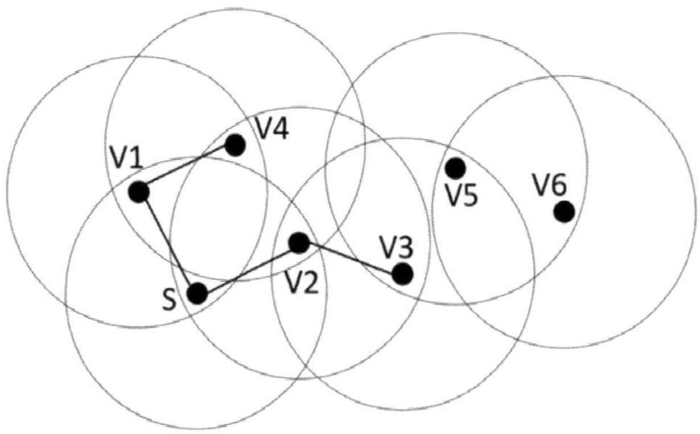

Figure 3. Links identified by the directly connected vehicles.

In the third step of the initial network scan, the server is using the same logic as in the previous step. This time, the CVi vehicles are those which were identified in the previous step and the MVj vehicles are, obviously, those not identified yet. For the above example, V5 will be found by V3.

This third step is repeated as long as there were vehicles identified in one iteration, so after the identification of V5, third step is repeated and V6 will be found by V5 (figure 4).

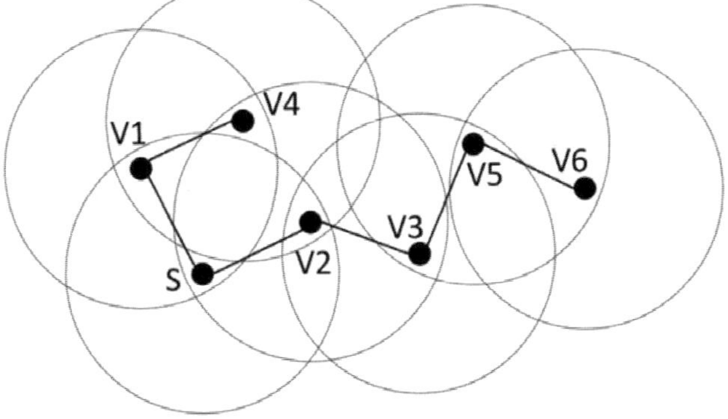

Figure 4. Links identified in the algorithm's third step.

After the initial network scan, the server will perform iterations for updating the data from each vehicle, based on a previously built connections graph. For each vehicle in the graph, the server will try first to reach the vehicle using the shortest path determined by the Dijkstra algorithm [10].

If the vehicle moved such as some of its graph connections are no longer valid, the server will try to contact it by computing the last known distance to other vehicles and using the estimated closest vehicle, except the one for which the connection was lost. For example, in figure 5, if vehicle V2 moved and its connection with the server was lost, the server will try to contact it using the path S – V1 – V4 – V2.

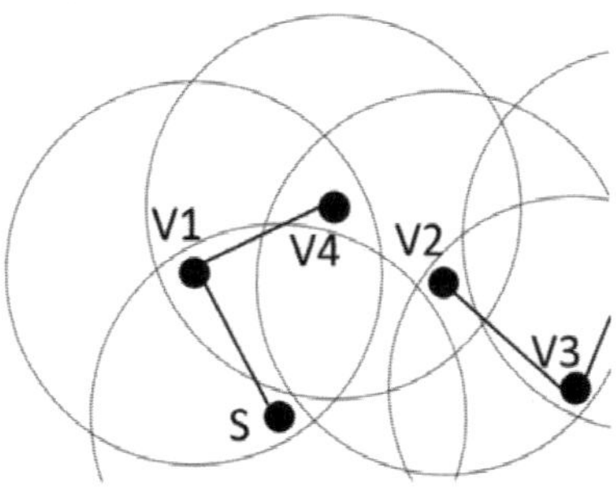

Figure 5. Path changing for connecting a lost vehicle.

If, at a certain moment, there are vehicles to which there is no path in the connections graph, the server will apply the steps from the initial network scan to try to reach them.

The algorithm was implemented in an application developed in the LabVIEW graphical programming environment [11].

4. Hardware testing

For both the server and the vehicles, the communication modules were designed using the HC-12 433MHz SI4463 wireless serial port module. The module is using the 433,4 - 473,0 MHz radio band and is allowing up to 5 Mbps data transmission rates over an up to 1 km distance in free air, using one hundred channels. The wireless modules were controlled using diverse types of Arduino development boards [12]. The server's development board is receiving the messages it has to send in the network from the LabVIEW application where the search algorithm was implemented, using an USB port.

Tests were first performed for assessing the coverage area, using 433 MHz, 5 dB antennas both on the server module and on one vehicle module which was placed in different areas, inside a building. The black circle in figure 6 represents the server, placed at the second floor. All other circles represent places at the ground floor: grey for the places where the communication was possible and white where there was no received signal.

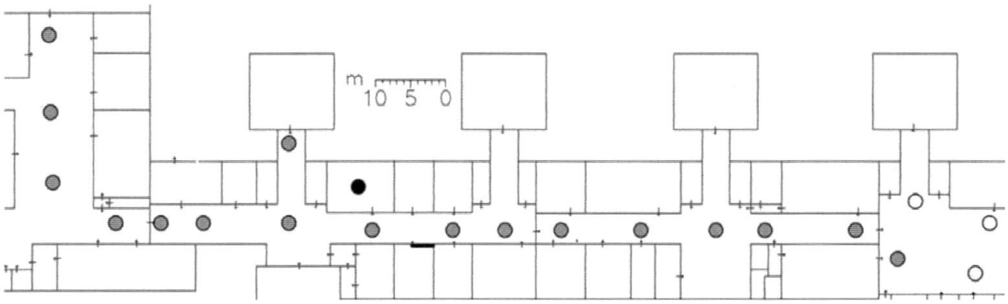

Figure 6. Coverage area for 5 dB antennas both on server and on vehicle.

A more limited coverage area was obtained when a short spring antenna was used for the vehicle module (figure 7).

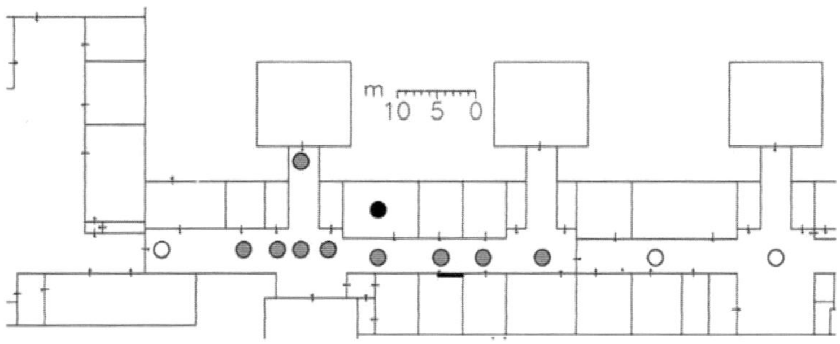

Figure 7. Coverage area for 5 dB antenna on the server and spring antenna on the vehicle.

Coverage area was extended, but not in a truly relevant way, when a fixed vehicle module with 5 dB antenna was placed in a room at the ground floor, acting as a relay for a second vehicle module with short spring antenna (figure 8). The couple of circles in the left part of the figure suggests that the signal reached the mobile module, via the fixed vehicle, but the return signal from the mobile module was not able to reach the fixed one, so the data didn't go back to the server.

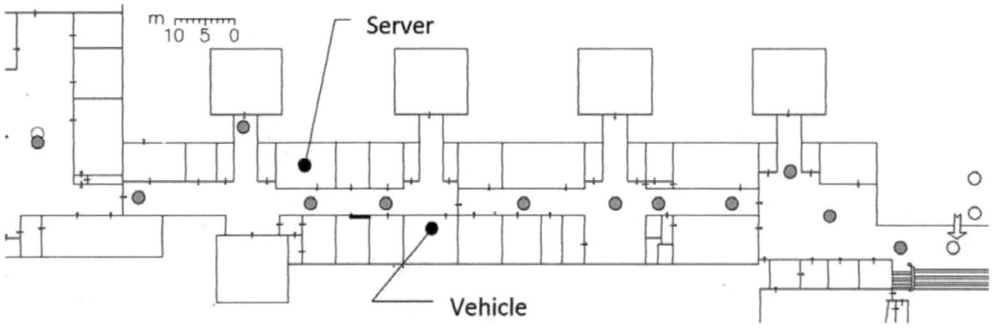

Figure 8. Coverage area when using a relay vehicle.

5. Conclusions

The implementation of the proposed communication algorithm proved to function correctly, the coverage areas obtained during the hardware testing being as one can expect. The algorithm can be useful not only in the situation mentioned in the introduction, but also when there is a chance for the communication network to fail and to affect the general functioning of the vehicles fleet. The fact that the algorithm performed well on the chosen development boards proved that the needed computing power is reduced, so the implementation can be used in applications were the hardware resources and the supplied power have to be carefully managed.

The effectiveness of the proposed search algorithm will depend on the total number of vehicles, on the number of unreachable vehicles and on how often the network structure is changing and is demanding the links to be redefined. A simulator will have to be developed for evaluating the algorithm's effectiveness in more complex situations.

Future developments will also address the dynamic situation when there is no server available, so the server's role will have to be played by one of the vehicles. The server role will be also able to be transferred from one vehicle to another. Criteria for activating the role switch will have to be established and a simulator will be used for testing the extended algorithm correctness, robustness and performance.

References

[1] Cristea M, Matache M G, Ioniță C, Perișoară L A, Cristea R D and Arsenoaia V N 2022 GSM WiFi Mobile Communication System for Agricultural Vehicles *INMATEH – Agr. Eng. J.* **67** 23-32

[2] Lei L and Wang Z 2021 *International Conference on Intelligent Computing, Automation and Systems (Chongqing, China)* IEEE p 198

[3] Ngoenriang N, Turner S J and Niyato D 2022 Joint UAV Placement and Data Delivery in Aerial Inspection Under Uncertainties *IEEE Internet of Things J.* **9** 6389-403

[4] Radzki G, Bocewicz G, Wikarek J, Nielsen P and Banaszak Z 2022 Comparison of exact and approximate approaches to UAVs mission contingency planning in dynamic environments *J. Math. Biosc. and Eng.* **19** 7091-121

[5] Lu S P, Lei C L, Ho C Y, Hwang S S and Chen H C 2022 Distributed Ledger Technology Based Architecture for Decentralized Device-to-Device Communication *Network IEEE Access J* **10** 92006-22

[6] Pratissoli F, Battilani N, Fantuzzi C and Sabattini L 2021 *IEEE International Conference on Robotics and Automation ICRA (Xian, China)* IEEE p 10009-15

[7] Chiu K L and Hwang R H 2012 Communication framework for vehicle ad hoc network on freeways *Telecomm. Sys. J.* **50** 243-56

[8] D'Amato E, Notaro I, Iodice B, Panico G, Blasi L 2022 *International Conference on Unmanned Aircraft Systems ICUAS (Dubrovnik, Croatia)* IEEE p 998-1005

[9] Zunino C, Valenzano A, Obermaisser R and Petersen S 2020 Factory communications at the dawn of the fourth industrial revolution *J. Comp. Stand. & Interf.* **71** 103433

[10] Blokdyk G 2018 *Dijkstra's Algorithm: End-to-end Data Analysis* (Create Space Independent Publishing Platform)

[11] Bitter R, Mohiuddin T and Nawrocki M 2017 *LabVIEW: Advanced Programming Techniques, Second Edition* (Boca Raton: CRC Press – Taylor & Francis Group)

[12] Javed A 2016 *Building Arduino projects for the Internet of Things: experiments with real-world applications* (New York: Apress Media)

IOP Conf. Series: Materials Science and Engineering 1268 (2022) 012008 doi:10.1088/1757-899X/1268/1/012008

Researches on the geometrical and dimensional accuracy for complex thin parts made of electrical steel by abrasive waterjet machining

Laurențiu-Aurel Mihail[1], Alexandru Filip[1], Adrian Mija[2]

[1] "Transilvania" University of Brașov, Bdul Eroilor nr.29, Romania

[2] Electroprecizia Electrical Motors Ltd., str.Electroprecizia nr.3, Săcele, Romania

E-mail: filipal@unitbv.ro

Abstract. The electrical steel laminations which compose the rotor and stator core for electrical motors have complex 2D shapes. When these parts require small batches of production, they can be efficiently manufactured only by flexible methods. The paper presents the results of using abrasive waterjet machining (AWJ) to obtain such parts within the frame of a research project with EU funding. The main objective of the research was the analysis of geometrical and dimensional accuracy of the parts which will confirm that AWJ is a suitable process in this case. The parts were measured on an optical digital profilometer. The measurement process was not simple because the parts are relatively large and cannot be focused only by one shot by the camera. Also, another important issue were the profile edges inconsistencies due to the burrs. Those problems and the limitations of the measurement technology did not allow the automation of measurement. The results and discussions proved that AWJ can be used successfully for manufacturing such parts and the machining process is traceable.

1. Introduction

Present and future electric motors which are widely used for the industrial equipment must comply the energy efficiency policies of the European Union. One of the requirements is a minimum increase of 20% for the motor's efficiency in the next years [1]. The IE3 generation of electrical motors will be replaced by the new classes IE 4 (Super premium efficiency) and IE5 (ultra-premium efficiency).

The design of IE4/IE5 electric motors is a very challenging task and sometimes require major changes in the motor structure and components, including new materials, shapes of laminations and new assembly methods [2]. These changes must be adopted with a minimum influence on the manufacturing process and costs. During the design process, a certain prototype of electric motor must be tested in terms of values of the functional parameters and energy efficiency, before being validated and approved for mass production.

Some of the most important components of an electric motor are the steel laminations which are stacked, welded or bonded in about a few hundred pieces to form the core of the stator and rotor. These parts are usually made of special electrical steels like M290-50A, M330-50A or M400-50A, according to DIN EN 10106 standard [3]. These materials assure mechanical and magnetic properties needed by the functional requirements of IE4 and IE5 motors.

The shape of the stator and rotor laminations is very complex and their thickness is very small, 0.5 or 0.65 mm. A typical pair of such parts is presented in figure 1. For large scale production these parts are made by stamping using a dedicated tool. For small scale production other manufacturing methods must be used, due to the high level of cost of the stamping tool. The processes which can be used in this case are laser cutting, abrasive waterjet cutting (AWJ) and wire electrical discharge machining (EDM).

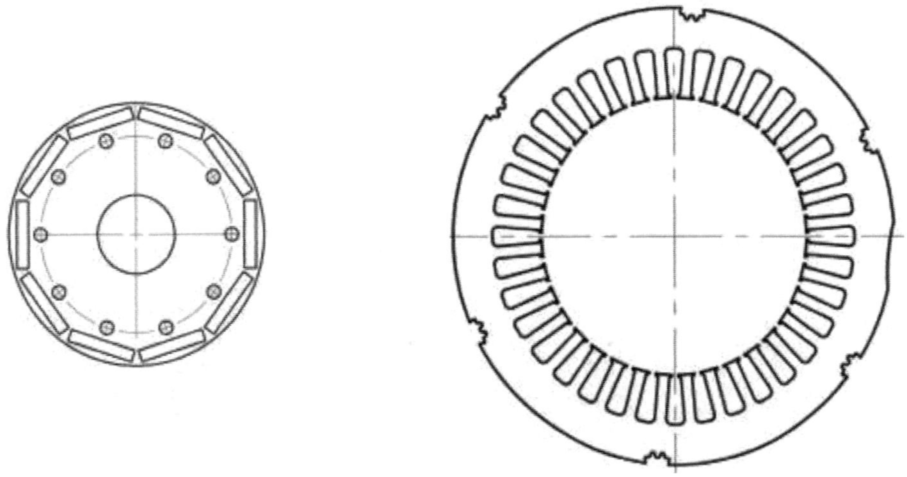

(a) Rotor lamination (b) Stator lamination

Figure 1. Typical shape of rotor and stator laminations (by courtesy of Electroprecizia Electrical Motors Ltd)

Researches made until today proved that the manufacturing process of stator and rotor steel laminations has a certain effect on the magnetic properties and even the chemical composition of the material. Peržel et al. [4] and Winter et al. [5] demonstrated that AWJ is the manufacturing method that has a minimum influence on the chemical and physical properties of the material and has no thermal influence on the cut zone, usually named HAZ (Heat Affected Zone). Paltanea et al [6] pointed out some correlations between the magnetic properties and the chemical composition of the electrical steels type M with the manufacturing process which was used for machining the parts. Schoppa et al [7] also came to same conclusion: AWJ is the process with a minimum influence on the material's properties.

All researches noticed that AWJ has some weak points like low manufacturing speeds, relatively long time setups and a relatively variable accuracy of the process in terms of dimensional precision. Fortunately, the first two disadvantages have a minimum importance in the case of small scale or prototypes production. The last one has been overcome in the last years by improving the precision of the AWJ head movements, but still the late researches [8, 9] mention this fact as a weak point of the process.

In previous researches the authors of the present paper showed that AWJ can be successfully used for machining parts made of special materials like Hardox [10] or special structures like aluminium honeycomb panels [11]. The dimensional precision of the parts was analysed and conclusions were made. The present research describes a measurement strategy, the challenges occurred and the geometrical and dimensional accuracy of a rotor lamination manufactured in a small scale production by abrasive waterjet cutting.

2. Experimental setup and measurements

The rotor lamination manufactured in a small scale production had the shape presented in Figure 1a.

The material of the part was M400-50A, an electrical steel with the properties and chemical composition according to DIN EN 10106 standard [3]. The thickness of the part was 0.5mm.

The process of abrasive waterjet cutting was performed on a Maxiem cutting machine which can reach a maximum water pressure of 350 MPa. The machine's dynamics does not have automatic tilt compensation and the abrasive feeding system is a regular one, having a constant feed rate. The parts were manufactured in sets of five pieces, placed on a strip of a certain length. For increasing the productivity of the process a pack of 3 strips were used for a simultaneous cutting. Figure 2 shows an image of the production system. The pack of strips was manually clamped on the traverse system of the machine table.

Some of the values of the technological parameters of the AWJ process were common ones, used by the authors in previous research. Other values like the traverse speed were adopted after a few preliminary tests. These values are presented in Table 1. Due to the complexity of the part's shape the traverse speed is variable along the machined contour, the value being automatically adjusted by the machine software, depending on the contour's curvature.

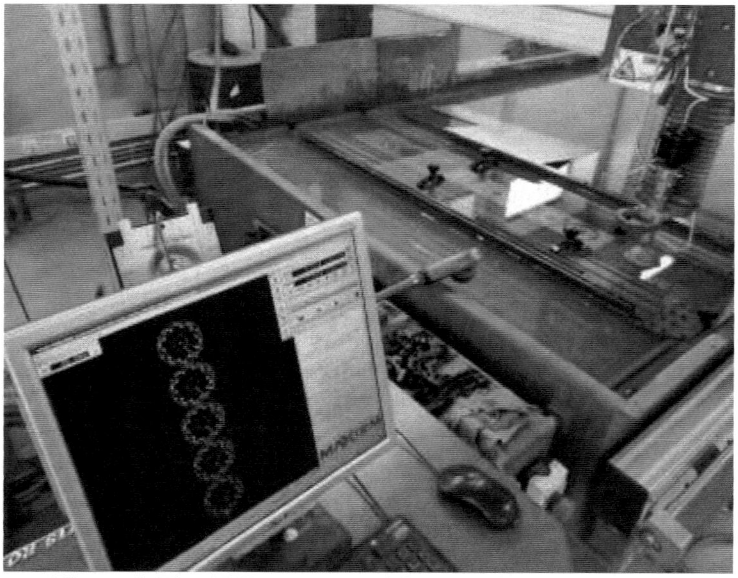

Figure 2. The AWJ system used for manufacturing the parts

Table 1. Values of the technological parameters used during the AWJ process.

Water pressure [bar]	Abrasive flow rate (g/min)	Abrasive type	Standoff distance (mm)	Traverse speed (medium) (mm/min)
3500	370	Garnet mesh 80	1.5	230

The measurements were performed on a digital profilometer, available in the "Metrology Lab Mitutoyo" from the "Institute High Tech Products for Sustainable Development: PRO-DD", lab supported by Mitutoyo Romania S.R.L.

The Quick Image digital profilometer, QI Vision, Figure 3, is a compact vision measuring system with the highest level of measuring accuracy in its class [12]. The sub-pixel image processing enables high-accuracy edge detection even with a wide view field, as well as stable, accurate measurements on relatively large parts. Its compact design and high-quality components make the QI Vision system ideal for integration directly on the shop floor or in the dimensional and geometry measurement laboratories.

The reliability of the digital profilometer consist on its accuracy and repeatability within the screen, both for high and normal resolution modes. The most impact point for its reliability is the edge identification features that are based on sub-pixel processing that enables a high-accuracy edge detection. The stage is highly accurate and very stable, allowing measurements for large parts. The telecentric optical system render the errors due to height to be minimized within a depth of focus with steps of up to 22mm and the measurement errors due to human focusing are eliminated. That feature is very useful for stepped parts and for revolution parts. Another aspect for the measurement's reliability is the calibration process with direct metrological traceability in correspondence with the national and international reference standard meter and the measurement environment.

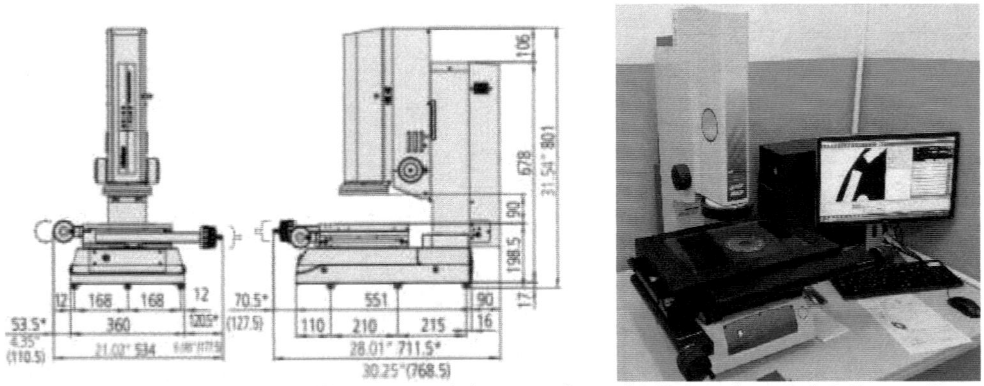

Figure 3. The digital profilometer Quick Image from the "Metrology Lab Mitutoyo"

The applications of a digital profilometer render is simple, because of the easy-to-use measurement facilities. The correction algorithm stitches multiple images together (multiple image-to-image coupling) while maintaining high-accuracy measurements. After a stitching operation, measurement is easily processed without the need to move to the next stage. The digital image zoom function is also very useful to maintain the measurement's accuracy. By the one-click execution function, after placing the workpiece within the field of view, the machine automatically recognizes the position and angle using a pattern search function and then finishes the measurements. There is no need for positioning and axially aligning the workpiece, the position and inclination of a workpiece can be measured even if it has moved. The feature can be recognized by a user defined configuration or even by the CAD drawing.

The efficiency in the measurement process of the digital profilometer is given by the measurement productivity. Even if the stage is not motorized, the profilometer is easy to use and quick to handle by the regular manual moving system or by the quick release moving system. The measurements are displayed on the screen very intuitively and the decision can be very quick if the N/NG function is applied. The stage surface allows the multiple parts layouts; therefore, the repeated measurement can be very accessible. The "Go/Not Go" (G/NG) feature can be applied, if the quality characteristics are defined by the tolerance fields. The high-definition color camera provides high-resolution color images for effective use in high-accuracy measurement and workpiece surface observation. Bright color measurement images can be easily stored as an image file and can be used for creating clear and understandable measurement report. The high-resolution mode produces the same wide field of view as the normal mode that operates with a deep focal depth. This allows to share a single measurement procedure for executing the measurements. The enhanced illumination function of the high-resolution mode enables measurements of low reflectivity workpieces like rubber and black resin moldings to be performed with a clear image. The edge detection functionality enables fast measurement by removing till a certain magnitude the outliers caused by anomalies such as debris, burrs and chips. The auto trace tool automatically detects the edges of unknown contours and obtains point group data if a supplementary software feature is available.

With a supplementary software, the statistical data can be displayed in real-time, making early detection of process irregularities possible. Early identification of an out-of-control situation enables rapid remedial action to be taken when necessary.

The focalization and the supplementary illumination that is crucial for the measurement are provided by the transmitted light - a green LED telecentric illumination, by the co-axial light - a white LED and by the ring light 4-quadrant white LED.

The QI Vision digital profilometer has the following certified accuracy values:

- The measurement accuracy within the screen for the high-resolution mode is ± 2 μm;
- The measurement accuracy within the screen for the normal resolution mode is ± 4 μm;
- The repeatability within the screen (± 2 σ) for the high-resolution mode is ± 1 μm;
- The repeatability within the screen (± 2 σ) for the normal resolution mode is ± 2 μm;
- The measurement accuracy (E1xy) is ± (3.5+0.02) μm, for an arbitrary measuring length (L, mm).

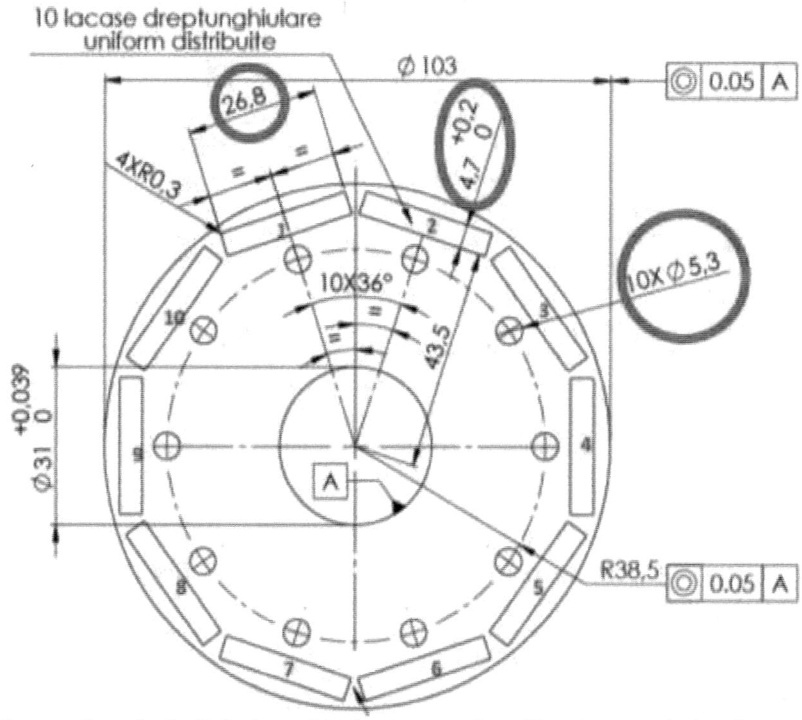

Figure 4. Measured part's draft design with the measured quality characteristics (by courtesy of Electroprecizia Electrical Motors Ltd)

For the rotor part the dimensional quality characteristics which were measured and analyzed, with specified tolerances and tolerances according to SR 2768 mK precision class [13], are presented in Table 2 and graphically highlighted in Figure 4.

Table 2. Part's measured dimensional nominal characteristics and it's tolerances.

	inner diameter D_{Int}	outer diameter D_{Ext}	circular hole, C_i	rectangular hole length, D_i	rectangular hole width, d_i
nominal value [mm]	31	103	5.3	26.8	4.7
upper tolerance [mm]	+0.039	+0.300	+0.1	+0.200	+0.200
lower tolerance [mm]	0.000	-0.300	-0.1	0.000	0.000

During the parts measurements there were encountered several issues because of the edges irregularities, resulted from the AWJM process. The edges state result to be irregular, with burs and various other types of quality issues, fact that makes impossible a reliable high precision & high accuracy automatic measurement, assisted by a repeatable part program. As consequence, all the 5 measured parts, for all the 32 quality characteristics measured on each one of them were ran manually. It is considered that the measurements are consistent because the initial equipment's setup was kept constant (focalization, saturation, automatic pixel selection properties, etc.) for all the 160 measurements developed for the studied part configuration. According with the production's study volume, the sample of 5 parts resulted to be a reliable one, taking into consideration the fact that the final process's outcome is a prototype development, with a very low production volume. Another important aspect on this topic is the fact that in the research and development project, previously, there were developed several comprehensive machinability tests for the abrasive water jet process, studies that are confidential for now, taking into consideration the overall project's output.

Figure 5 presents the user interface of the programming and operation QI Pack software, used for mastering the digital profilometer during the part measurement specific activities.

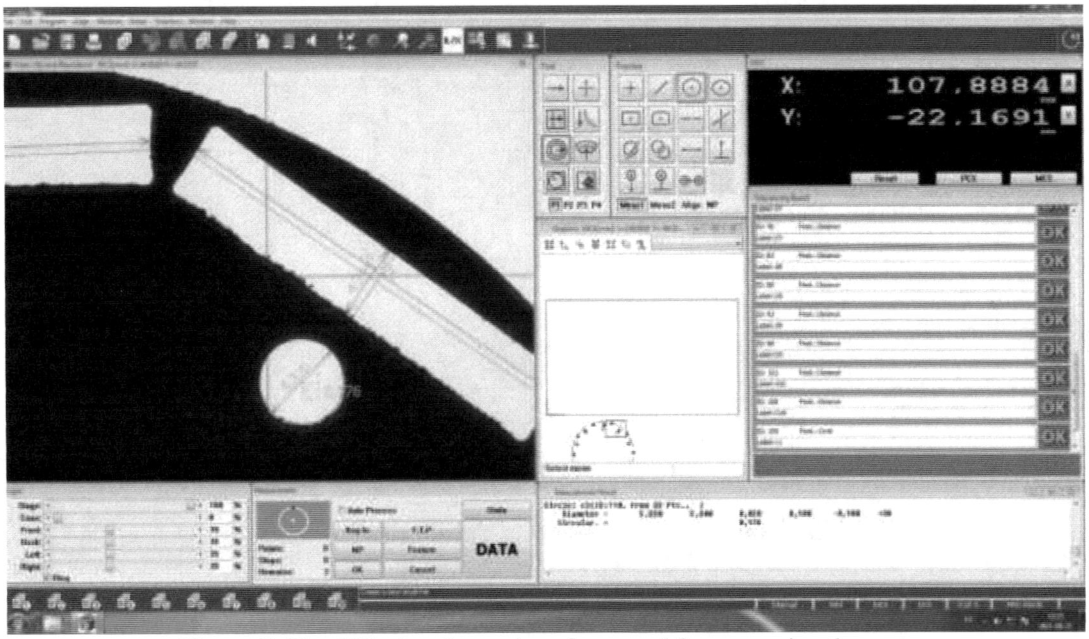

Figure 5. The QIPack software interface, used for measuring the parts

3. Analysis of the parts' geometrical and dimensional accuracy

The dimensional measurement data for the rectangular hole width, [mm] measured on the digital profilometer is presented in Table 3. By applying the outlier Grubbs's test, there were encountered 2 outliers, the fourth rectangular hole width of the part 1 and the fifth rectangular hole width of the part 3.

The method applied considered the null hypothesis as all data values come from the same normal population and the alternative hypothesis, smallest or largest data value is an outlier, for a significance level $\alpha = 0.05$.

The output of the Grubbs's test is presented in the Figure 6, encountering the smallest *P* statistics for the part 1 and the part 3, with corresponding *G* largest values that indicate the places of the outliers, marked in Table 3 and represented in the figure 7.

IOP Conf. Series: Materials Science and Engineering 1268 (2022) 012008 doi:10.1088/1757-899X/1268/1/012008

Table 3. Dimensional measurement data for the rectangular hole width [mm].

	Part 1	Part 2	Part 3	Part 4	Part 5
Rectangular hole 1	4.785	4.686	4.930	4.674	4.808
Rectangular hole 2	4.744	4.720	4.901	4.620	4.727
Rectangular hole 3	4.747	4.720	4.901	4.620	4.727
Rectangular hole 4	4.647	4.678	4.884	4.594	4.780
Rectangular hole 5	4.816	4.680	4.760	4.659	4.812
Rectangular hole 6	4.810	4.707	4.912	4.663	4.812
Rectangular hole 7	4.824	4.676	4.893	4.645	4.696
Rectangular hole 8	4.841	4.698	4.898	4.652	4.670
Rectangular hole 9	4.796	4.678	4.937	4.593	4.688
Rectangular hole 10	4.806	4.687	4.866	4.640	4.803

Grubbs' Test

Variable	N	Mean	StDev	Min	Max	G	P
Part 1	10	4.7816	0.0566	4.6470	4.8410	2.38	0.026
Part 2	10	4.6930	0.0172	4.6760	4.7200	1.57	0.991
Part 3	10	4.8882	0.0495	4.7600	4.9370	2.59	0.003
Part 4	10	4.6360	0.0282	4.5930	4.6740	1.52	1.000
Part 5	10	4.7523	0.0567	4.6700	4.8120	1.45	1.000

Figure 6. The Grubbs's test output, applied for the measurement data for the rectangular hole width

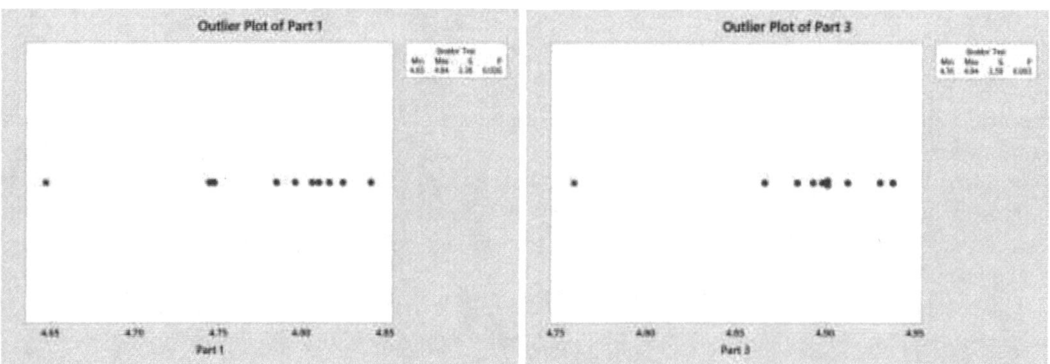

Figure 7. The Grubbs's test output, applied for the measurement data for the rectangular hole width

Relatively to the specified tolerance limits, the measurements of the rectangular holes widths result to be entirely nonconform for the part 4 and with several nonconform outputs for the part 2, relatively at the inferior tolerance limit. Relatively to the superior tolerance limit, only the part 3 present three nonconform outputs (see Figure 8).

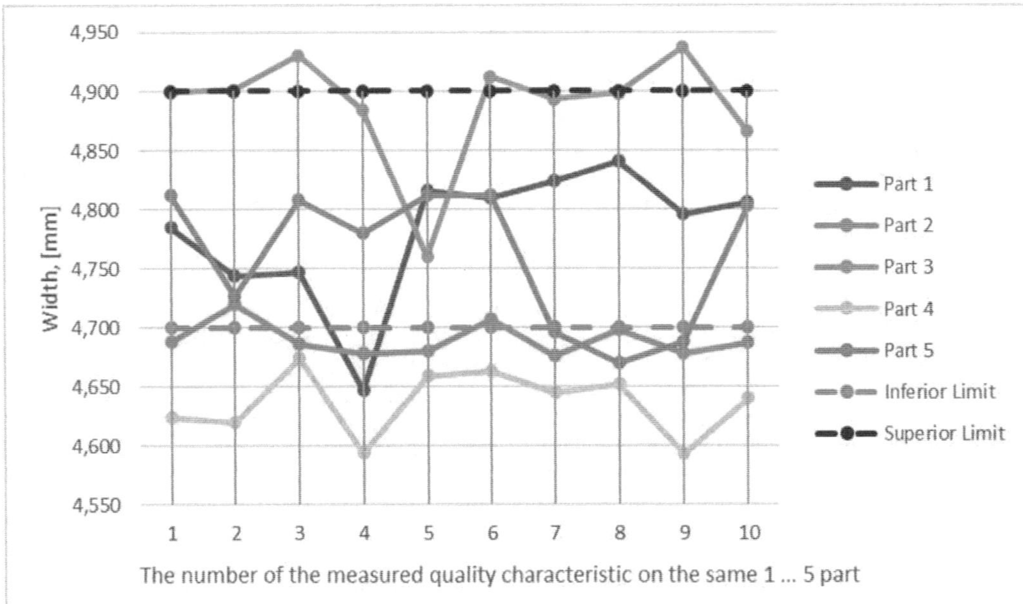

Figure 8. The measurements data representation for the width quality characteristic relatively to the specified tolerance limits

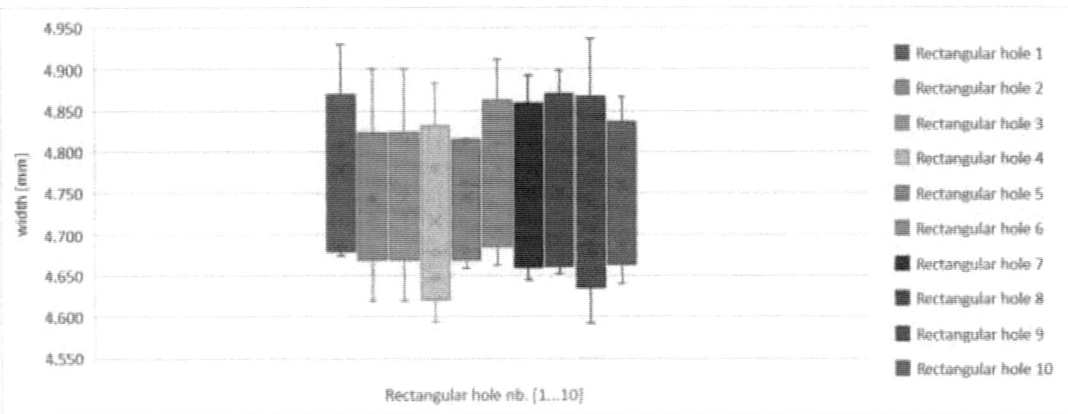

Figure 9. The boxplot diagram for the measurements of the width quality characteristic

The boxplot representation (Figure 9) for the measurements of the rectangular hole width quality characteristic is used for the final representation of the data analyzed, in order to assess and compare the shape, central tendency, and variability of sample distributions. The outliers are already identified by the previously presented Grubbs's test. The boxplot shows the median, interquartile range, the data range, and outliers for each group of data (the 10 measurements for each of the 5 parts apart). The median is represented by the line in the box, being a common measure of the center of the data; half the observations are less than or equal to it, and half are greater than or equal to it. The interquartile range

IOP Conf. Series: Materials Science and Engineering 1268 (2022) 012008 doi:10.1088/1757-899X/1268/1/012008

box represents the middle 50% of the data, showing the distance between the first and third quartiles (Q3-Q1). The whiskers extend from either side of the box representing the ranges for the bottom 25% and the top 25% of the data values, excluding the encountered outliers. Even if the sample of data can be considered a relatively low one (5 parts), the overall representation results to be reliable. In terms of skewness, the part 1 and the part 5 results to be right, respectively left skewed, that suggest that the two parts measurements are susceptible not to be normally distributed. The Anderson Darling statistic for the first hole width (5 parts) is 0.274 and the p value is 0.498. For the fifth hole width (5 parts), the Anderson Darling statistic is 0.355 and the p value is 0.291.

Table 4 presents the dimensional measurement data for the rectangular hole length, [mm], graphically represented within the Figure 11, relatively at their specified tolerance limits. By applying the outlier Grubbs's test, there were not encountered outliers. The method applied took, also, into consideration the null hypothesis as all data values come from the same normal population and the alternative hypothesis, smallest or largest data value is an outlier, for a significance level $\alpha = 0.05$. The output of the Grubbs's test is presented in the Figure 10.

Table 4. Dimensional measurement data for the rectangular hole length [mm].

	Part 1	Part 2	Part 3	Part 4	Part 5
Rectangular hole 1	26.889	26.817	27.093	26.831	26.998
Rectangular hole 2	26.926	26.789	27.044	26.778	26.995
Rectangular hole 3	26.920	26.820	27.039	26.818	27.024
Rectangular hole 4	26.874	26.805	27.034	26.777	26.981
Rectangular hole 5	27.006	26.876	27.110	26.859	27.037
Rectangular hole 6	26.916	26.794	27.033	26.777	26.980
Rectangular hole 7	27.000	26.794	27.012	26.781	26.972
Rectangular hole 8	26.977	26.813	27.058	26.728	26.974
Rectangular hole 9	26.956	26.721	27.041	26.835	26.982
Rectangular hole 10	26.933	26.888	27.096	26.738	27.035

Grubbs' Test

Variable	N	Mean	StDev	Min	Max	G	P
Part 1	10	26.940	0.0444	26.874	27.006	1.49	1.000
Part 2	10	26.812	0.0465	26.721	26.888	1.95	0.288
Part 3	10	27.056	0.0325	27.012	27.110	1.66	0.762
Part 4	10	26.792	0.0426	26.728	26.859	1.57	0.986
Part 5	10	26.998	0.0252	26.972	27.037	1.56	1.000

Figure 10. The Grubbs's test output, applied for the measurement data for the rectangular hole length

According with Figure 12, in terms of skewness, the part 1 and the part 3 result to be right skewed, suggesting the fact that the measurements are susceptible not to be normally distributed. The Anderson Darling statistic for the first hole length (5 parts) is 0.295 and the p value is 0.436. For the third hole length (5 parts), the Anderson Darling statistic is 0.384 and the p value is 0.238.

Table 5 presents the dimensional measurement data for the circular hole diameter, [mm], graphically represented in Figure 15. By applying the outlier Grubbs's test, it was encountered 1 outlier, the sixth circular hole width of the part 3. The method applied took, also, into consideration the null hypothesis as all data values come from the same normal population and the alternative hypothesis, smallest or largest data value is an outlier, for a significance level $\alpha = 0.05$.

The output of the Grubbs's test is presented in Figure 13, encountering the smallest P statistic for the part 3, with corresponding G largest value that indicate the places of the outlier, marked in Table 5 and represented in Figure 14.

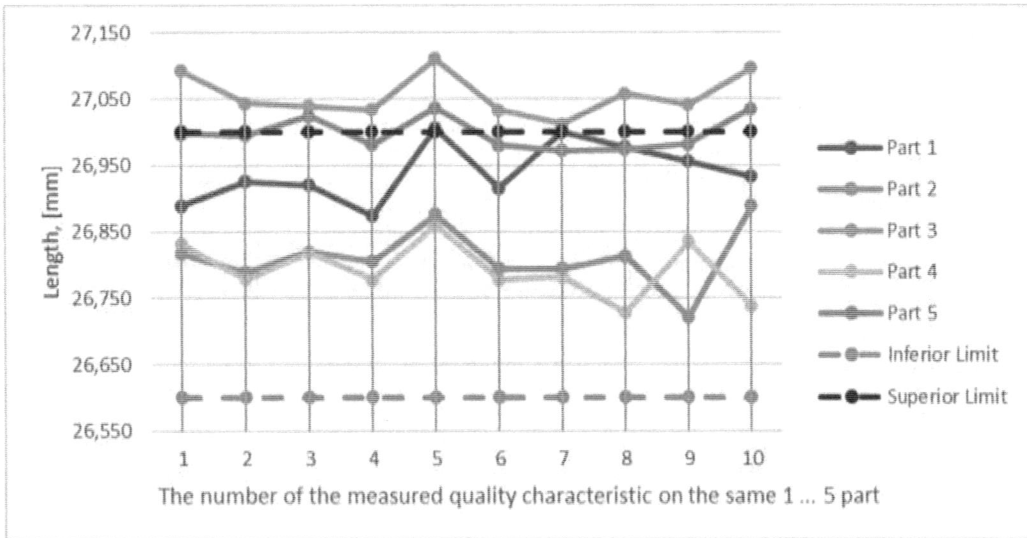

Figure 11. The measurements data representation for the length quality characteristic relatively to the specified tolerance limits

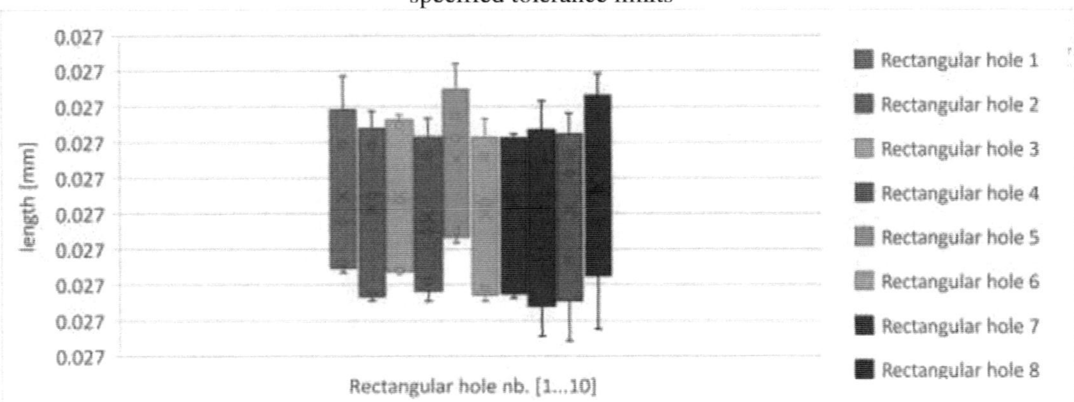

Figure 12. The boxplot diagram for the measurements of the length quality characteristic

Grubbs' Test

Variable	N	Mean	StDev	Min	Max	G	P
Part 1	10	5.3352	0.0117	5.3200	5.3550	1.69	0.698
Part 2	10	5.2387	0.0170	5.2120	5.2730	2.02	0.217
Part 3	10	5.4610	0.0111	5.4450	5.4870	2.34	0.035
Part 4	10	5.2265	0.0128	5.1990	5.2470	2.14	0.121
Part 5	10	5.3260	0.0225	5.2900	5.3610	1.60	0.909

Figure 13. The Grubbs's test output, applied for the measurement data for the circular hole diameter

Table 5. Dimensional measurement data for the circular hole diameter [mm].

	Part 1	Part 2	Part 3	Part 4	Part 5
Circular hole 1	5.327	5.242	5.445	5.223	5.350
Circular hole 2	5.320	5.212	5.465	5.228	5.336
Circular hole 3	5.331	5.273	5.461	5.236	5.317
Circular hole 4	5.334	5.225	5.450	5.234	5.333
Circular hole 5	5.322	5.241	5.460	5.231	5.328
Circular hole 6	5.333	5.243	5.487	5.247	5.361
Circular hole 7	5.340	5.232	5.465	5.199	5.326
Circular hole 8	5.353	5.256	5.460	5.229	5.290
Circular hole 9	5.355	5.236	5.461	5.217	5.291
Circular hole 10	5.337	5.227	5.456	5.221	5.328

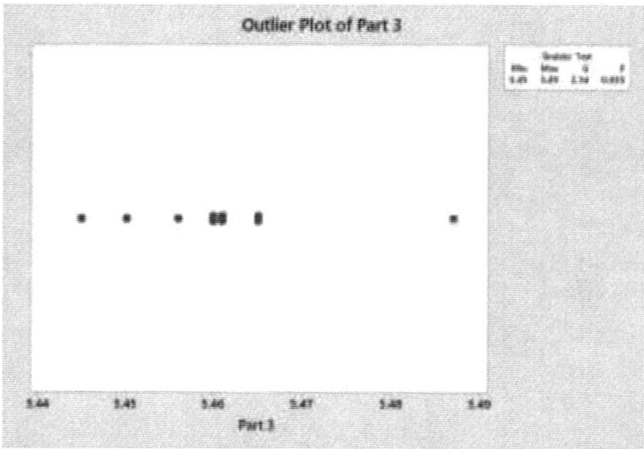

Figure 14. The Grubbs's test output, applied for the measurement data for the circular hole diameter

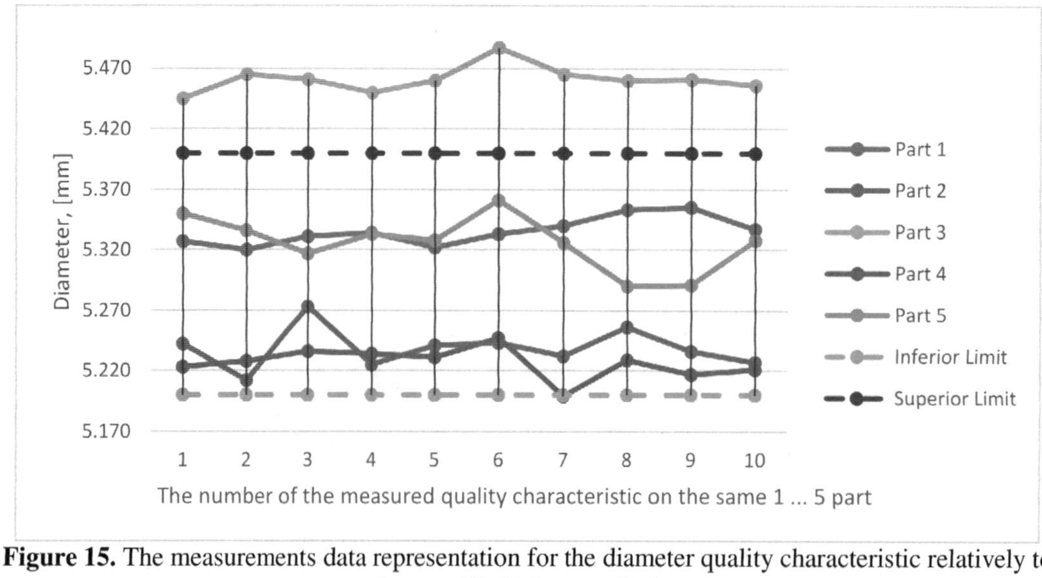

Figure 15. The measurements data representation for the diameter quality characteristic relatively to the specified tolerance limits

IOP Conf. Series: Materials Science and Engineering 1268 (2022) 012008 doi:10.1088/1757-899X/1268/1/012008

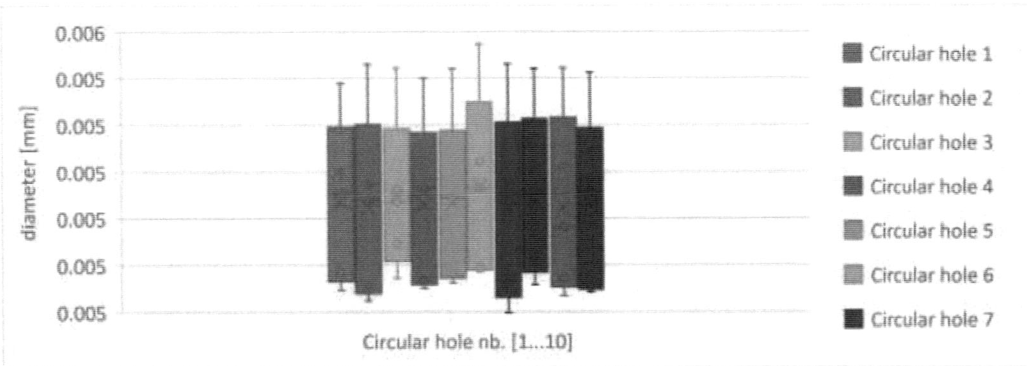

Figure 16. The boxplot diagram for the measurements of the diameter quality characteristic

According with Figure 16, in terms of skewness all the parts result to be right skewed, indicating that the five parts measurements are susceptible not to be normally distributed for all the 10 diameters. The Anderson Darling statistic was computed for all the parts and is presented in Table 6.

Table 6. *G* and *p-values* statistics of measurement data for the circular hole diameter [mm].

Diameter 1	Diameter 2	Diameter 3	Diameter 4	Diameter 5
G statistic	*G statistic*	*G statistic*	*G statistic*	*G statistic*
p-value	*p-value*	*p-value*	*p-value*	*p-value*
0.234	0.284	0.308	0.301	0.347
0.610	0.468	0.399	0.342	0.307
Diameter 6	Diameter 7	Diameter 8	Diameter 9	Diameter 10
G statistic	*G statistic*	*G statistic*	*G statistic*	*G statistic*
p-value	*p-value*	*p-value*	*p-value*	*p-value*
0.307	0.239	0.271	0.251	0.320
0.403	0.593	0.507	0.553	0.369

4. Conclusions

For the rectangular holes, machined by abrasive water jet technology and circularly disposed on the part, it can be pointed out the following conclusions:

- The linear edges, cut by abrasive water jet of the electrical steel laminations can be measured with a superior accuracy along their length;
- Along the small rectangle's hole edge, the number of points (pixels) that can be selected is much lower, therefore, the edge cannot be very accurately approximated by the measuring equipment, resulting higher measurement errors;
- Along the long edge of the rectangle it can be selected much more points (pixels) in order to approximate the measured edge, therefore the measurements accuracy is higher;
- According with the already specified phenomenon, at the previous points, can be judged the fact that the parts 2 and 4 result to present nonconformities in a consequent manner on the width side and, meanwhile, they are conform on the length quality characteristic direction;
- All the measurements result to be consequent relatively to the measured parts, putting in evidence that the measuring process was developed in a reliable way;
- The length measurements results reveal a process that is situated outmost toward the superior tolerance limit.

For the circular holes, circularly disposed on the measured parts, it can be underlined the following conclusions:

- It exists an obvious consistency between the part 1 and part 5 measurements, respectively the parts 2 and 4 measurements. The part 3 result to be nonconform, in a sustainable way, over the superior tolerance limit;
- The randomness and the disposal relatively to the specified tolerance limits of the measurements, and their consistency for all the measured circular holes (10 on each of the 5 parts) is because of the machining process stability for the circular interpolation.

The overall measurement results show that the parts are good for use, even there are some nonconformities according to the dimensional requirements. The above research proved that machining complex thin parts by abrasive waterjet is a challenging task. The measurement method applied, by using a digital profilometer, is a reliable one and it proved to be suitable for this type of parts, considering the low volume production rate. The method is accurate, easy-to-use and can be automated for higher batches of measurements, for an increase of the productivity.

Acknowledgments
The experiments of this research were supported by the project ”Creşterea competitivităţii Electroprecizia Electrical Motors prin dezvoltarea în parteneriat cu Universitatea Transilvania - Braşov a unei noi familii de motoare electrice, cu eficienţă energetică de clasă superpremium (IE4)/121228/POC/163/1/3”, co-financed by EU.

References

[1] Ferrara R 2015 *The Smart City and the Green Economy in Europe: A Critical Approach Energies* 2015 8 p 4724–4734.
[2] Cavagnino A, Vaschetto S, Ferraris L, Gmyrek Z, Agamloh E and Bramerdorfer G 2018 *Towards an IE4 Efficiency Class for Induction Motors with Minimal Manufacturer Impact IEEE Energy Conversion Congress and Exposition (ECCE)* p 289-296
[3] Available at https://www.en-standard.eu/ Cold rolled non-oriented electrical steel strip and sheet delivered in the fully processed state DIN EN 10106
[4] Peržel V, Hloch S, Tozan H, Yagimli M and Hreha P 2011 *Comparative analysis of abrasive waterjet (AWJ) technology with selected unconventional manufacturing processes Int J of the Physical Sciences* Vol 6(24), p 5587-5593
[5] Winter K, Liao Z, Ramanathan R, Axinte D, Vakil G and Gerada C 2021 *How non-conventional machining affects the surface integrity and magnetic properties of non-oriented electrical steel Materials & Design* Vol 210 110051
[6] Paltanea G, Manescu V, Stefanoiu R, Nemoianu I V and Gavrila H 2020 *Correlation between Magnetic Properties and Chemical Composition of Non-Oriented Electrical Steels Cut through Different Technologies Materials* 13 (6) 1455.
[7] Schoppa A, Louis H and Pude F 2003 *Influence of abrasive waterjet cutting on the magnetic properties of non-oriented electrical steels J Magn Magn Mater* 254: 370–372
[8] Saravanan S, Vijayan V, Jaya Suthahar ST, Balan AV, Sankar S and Ravichandran M 2020 *A review on recent progresses in machining methods based on abrasive water jet machining Materials Today: Proceedings* Vol 21 Part 1 p 116-122
[9] Llanto JM, Tolouei-Rad M, Vafadar A, Aamir M 2021 *Recent Progress Trend on Abrasive Waterjet Cutting of Metallic Materials: A Review Applied Sciences* 11 (8):3344
[10] Bulea H, Paunescu R and Filip AC 2015 *Dimensional Variation of Aluminum Honeycomb Panel on Circular Cutting with Abrasive Jet Machining Technology Applied Mechanics and Materials*, vol 760 Trans Tech Publications Ltd. p 397
[11] Filip AC, Mihail LA and Vasiloni MA 2017 *An experimental study on the dimensional accuracy of holes made by abrasive waterjet machining of Hardox steels MATEC Web Conf.* 137 02003.
[12] Mitutoyo technical documentation (2018) *Mitutoyo Quality – 2D Color Vision Measuring System Quick Image Series*, Bulletin number 2186(2), Mitutoyo America Corporation
[13] ISO 2768-1:1989 (EN) *General tolerances — Part 1: Tolerances for linear and angular dimensions without individual tolerance indications*, International Standards Organization

IOP Conf. Series: Materials Science and Engineering 1268 (2022) 012009 doi:10.1088/1757-899X/1268/1/012009

Interferometric evaluation of positioning error and position repeatability on CNCs

F P G Stochioiu[1,2], C Stochioiu[2*], A M E Rolea[1] and N Gannam[2,3]

[1] National Institute of Research and Development in Mechatronics and Measurement Technique Bucharest (INCDMTM), Şoseaua Pantelimon 6-8, Bucharest, Romania, 021631
[2] University POLITEHNICA of Bucharest, Splaiul Independenței 313, Bucharest, Romania, 060042
[3] Dr. Köcher S.R.L., Baicului Street 82A, Bucharest, Romania, 021784

E-mail: c.stochioiu@gmail.com

Abstract. The interferometer is widely used for various applications concerning high precision distance and position measurements, especially for the calibration of numerically controlled machines. While the setup procedures of the interferometer may be provided by the manufacturer, some calibration procedures are well defined by the international standards. Since the operation of interferometers requires qualified personnel, some helpful hints for the novice users are to be presented in this paper. Even if the interferometer is highly recommended by international calibration standards, some factors are prone to intervene into the quality of results. This paper presents the experimental results for the evaluation of positioning accuracy and repeatability on Y axis of two vertical machining centres. The results show the high importance of environmental conditions compensation, the impact of time pauses during the measurement procedure and the implications of the increment of measuring target positions.

1. Introduction

Presently, the interferometer is one of the most precise measuring instruments for distance and positioning [1]. The first devices using the principle of interferometry appeared in the 19th century and developed since, with numerous current applications, such as identifying a new method of defining the meter [2], plasma diagnostics [3], precision industrial measurement of lengths and movement [4,5], linear deviations and strain measurements [6], gravitational antennas [7], vibration analysis [8], illegal mining and landslides monitoring [9, 10], blood speed measurement [11], diseased cells detection [12], etc.

Laser interferometers are widely used for the calibration of fabrication equipment, due to their high accuracy and reliability [13]. Generally, an unconstrained part has 6 degrees of freedom, being able to move along or rotate around any of the orthogonal axes. In the literature a total of 18 deviations from these movements are defined, six by each axis, to which the squareness errors between the axes are added. Therefore, up to 21 errors can appear in a moving component. These errors are also found in three-dimensional working NC manufacturing machines and CMMs. Out of these 21 errors, with the help of interferometric devices, one can perform measurements for linear and angular positioning, straightness, flatness, squareness and parallelism evaluations for all the movements of the component. [14, 15]

Along with the advantages of using interferometers, there are also some uncertainties and limitations of this measuring method, which need to be overcome in order to obtain a precise and unaltered measurement. One of the most encountered issues is the environmental conditions instability,

IOP Conf. Series: Materials Science and Engineering 1268 (2022) 012009 doi:10.1088/1757-899X/1268/1/012009

the laser beam requiring a stable environment in order to obtain reliable results. Furthermore, the set-up of the optical elements might determine the measuring range to not cover the whole length of the evaluated axis, the increment selected for the measuring targets might affect the quality of the results. The whole set-up procedure, realignment of the laser beam and the conduct of the procedure is time consuming and requires qualified personnel, leading to relatively high costs for a calibration procedure.

It is of great significance to determine the impact of these factors since they can influence the quality of the measuring results. This study is focused on some of these factors that might intervene into the measuring process and can have a negative influence.

2. Materials and methods

Often, the interferometer manufacturers provide a setup procedure for each measurement type, while various international standards elaborate the procedures to be performed according to the tested characteristic of the machine. For this study, the ISO 230-2 [16] standard is to be used. This standard describes the measuring procedure for accuracy evaluation of numerically controlled machines, especially for evaluation of their positioning accuracy and repeatability on each linear or circular axis. The outcome of these measurements can be unidirectional or bidirectional values of positioning errors, positioning repeatability and systematic positioning errors, reversal errors, mean reversal errors, range of mean bidirectional positioning errors, etc.

The machines to be evaluated are two vertical machining centres, a Makino D200 Z and a Spinner MC 1020. The Makino D200 Z machining centre has five axes of movement, one linear and two circular for the working table, and two linear for the spindle. The Spinner MC 1020 machining centre has three axes of movement, all on the spindle.

The interferometer used for the experimental part of this paper is a Renishaw XL 80 with an environmental compensation unit XC-80. This interferometer is relatively easy to set up, in comparison with other commercially available interferometers. The setup for a positioning accuracy and repeatability evaluation for Y-axis on a CNC is represented in figure 1 and its procedure is as follows:

- The laser head, which is also the interference receptor, is placed on a tripod and adjusted with a bubble head; the laser head is connected to a laptop;
- The laptop is equipped with a software that displays in real time the results and captures the measured values. For the present work, a proprietary Renishaw software was used, called Renishaw Carto - Capture;
- The laptop is also connected to an environmental compensation unit, which has two sensors, for room atmosphere and machine temperature, placed in the measuring area;
- The alignment procedure begins with a positioning of one of the first lens assembly, the reflector, to intersect the beams' path and create its return path to the receptor situated into the laser head; the reflector is then moved along the whole measuring range of the axis to be evaluated and is aligned as many times as needed along the axis in order to keep the alignment along the whole measuring range;
- The second lens assembly, formed with a beam splitter and a reflector is then placed along the path of the beam, on a different part than the initial splitter, and aligned as required;
- The measuring process begins at the closest point between the two lens assemblies, and it is conducted as suggested by the standards.

The Renishaw interferometer is aligned for Y axis positioning accuracy and repeatability evaluation of the spindle of a CNC D200 Z vertical machining centre, as can be seen in figure 1. The lines represented with red and orange, are marking the initial path and the returning path respectively. Depending on the moving element, the initial beam path uses one reflector for reference while the secondary reflector travels along the measuring area. The interference occurs on the return path, after the moving beam interfere constructively and destructively with the reference beam when their path reunites by means of the same beam splitter lens forming a series of fringes. Positioning is measured by counting the number of fringes produced by the movement of the reflector.

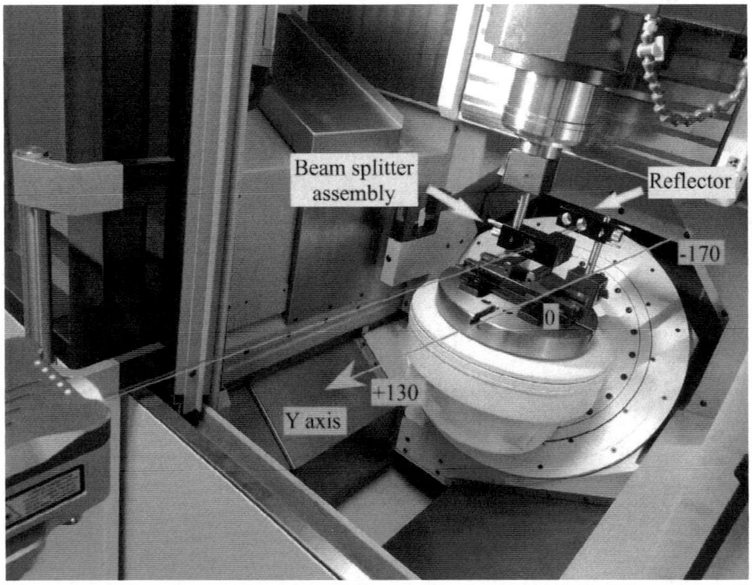

Figure 1. General setup of a Renishaw XL 80 interferometer on the Makino D200 Z vertical machining centre, for positioning accuracy evaluation on Y axis

Some procedural elements to be taken into consideration are:

- The interferometer components have to be soaked into the same environment condition as the CNC to be evaluated in order to avoid uncertainties determined by the material thermal dilation;
- The reflector must be aligned first, along the whole axis travel;
- The beam splitter has to be positioned on a rigid, fixed component, allowing a measuring range as large as possible;
- The first measuring point along the axis has to be as close as possible to the beam splitter, in order to avoid dead path error [17];
- A minimum of five cycles are recommended for one test;
- The laser source might overheat, therefore well-planned measuring tests are recommended.

In the present paper, a linear axis is to be evaluated on the positioning accuracy and repeatability, in correlation with ISO 230-2. Therefore, the measuring area on the selected axis to be evaluated will have several measurement points where the interferometer will bidirectionally determine the positioning error.

3. Experimental analysis

The measuring quality is given by various factors such as the operator skill level, environmental conditions, unexpected procedure interruptions, measuring increments etc. Some of these factors were experimentally analysed in order to determine their influence.

3.1. The importance of environmental auto compensation

The effects of interferometer compensation algorithms are put into perspective, with the help of two tests: one with the auto compensation activated unit and one without it. As the manufacturer states, the XC-80 unit is able to capture and feed the software with the environmental conditions, while the software auto compensates the effects on the laser beam wavelength. According to the environmental conditions captured by the two sensors, the auto compensator modifies the wavelength's nominal value with the true value, which is then used in calculation of the length, eliminating any measuring errors caused by the environment.

The two ISO 230-2 measurements were performed on the Y axis displacement for the spindle of a CNC D200Z machining centre. They are represented in figure 2. Along the abscissa, the Y axis range values are kept to [+10 mm ÷ -170 mm], where the positive values are directed in the sense of the Y axis shown in figure 1. The standard and the manufacturer recommend an environmental temperature of 20° C [16]. The environmental conditions in these tests are of around 16° C air temperature and of around 15° C material temperature as a means to evaluate the impact. It is worth noting that the CNC machine used for the test is a research facility unit, with little to no wear. As such, positioning errors are expected to be low, close to the ones presented by the producer of the machine.

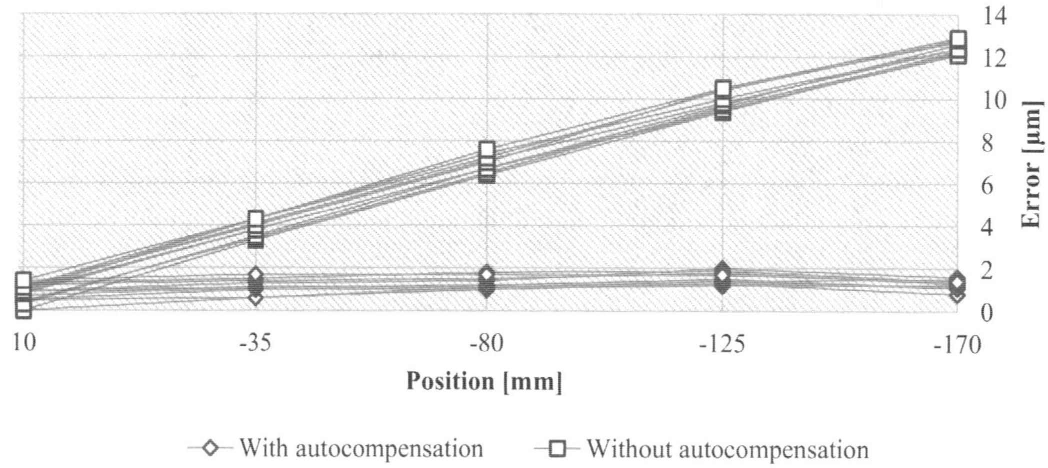

Figure 2. Y axis accuracy and repeatability of positioning measurements with and without auto compensation

A rather significant difference between the values given by the measurement with auto compensation, where the results show little to no error, and the values given by the measurement without auto compensation, where the error goes up to over 12 µm. The results of the two measurements are gathered in table 1, where it can be observed that, while the positioning repeatability is almost the same value, the positioning errors have a difference of around 11 µm. For the two measurements, the environmental conditions varied with less than 0,5° C for the temperature while the other conditions were stable.

The tests of evaluating the XC-80 capacity demonstrated the large effect of the compensation algorithms over the results even with small condition variations which lead to wavelength changes.

Table 1. Mean values of linear positioning evaluations with/without auto compensation, on Y axis

Measurement	Name / Direction	Positioning error	Positioning repeatability	Systematic positioning error	Range of mean bidirectional positioning error
With auto compensation	(+) [µm]	1,7	1,5	0,6	-
	(-) [µm]	2,6	2,0	0,9	-
	Bidirectional [µm]	2,6	2,0	0,9	0,7
Without auto compensation	(+) [µm]	12,9	1,9	11,5	-
	(-) [µm]	13,4	2,1	11,9	-
	Bidirectional [µm]	13,4	2,1	11,9	11,7

3.2. Time dependent results

ISO 230-2 recommends a continuous, programmed procedure for each measuring set. Therefore, an interrupted measurement might be considered a failure. The main reason for this is the environmental instabilities. Unwanted interruptions might intervene at any moment, due to human or other errors. The effects are to be analysed, by imposing an interruption of several minutes in the measurement cycle. As seen in the previous paragraph, slight changes in environmental conditions risk of affecting the measurement, if no wavelength compensation is applied. For the present test, the automatic compensation will be used, facilitated by the XC-80 environmental compensation unit.

This type of measurement pause might appear during an on-site measurement, where the unconstrained work conditions of a laboratory cannot be guaranteed, as is the case of production facilities, for example.

An experimental procedure at an auto compensated, steady 28° C air temperature, with 1008,4 mbar air pressure, 43 % RH air humidity and 26,4° C material temperature was performed on the SPINNER MC 1020 machining centre, in order to determine the time dependence of the results. The measuring process was stopped for around 5 minutes during the third cycle of the test, as seen in figure 3. It is worth noting that the machine used for the experiments has been extensively used in production and non-negligible positioning errors are expected.

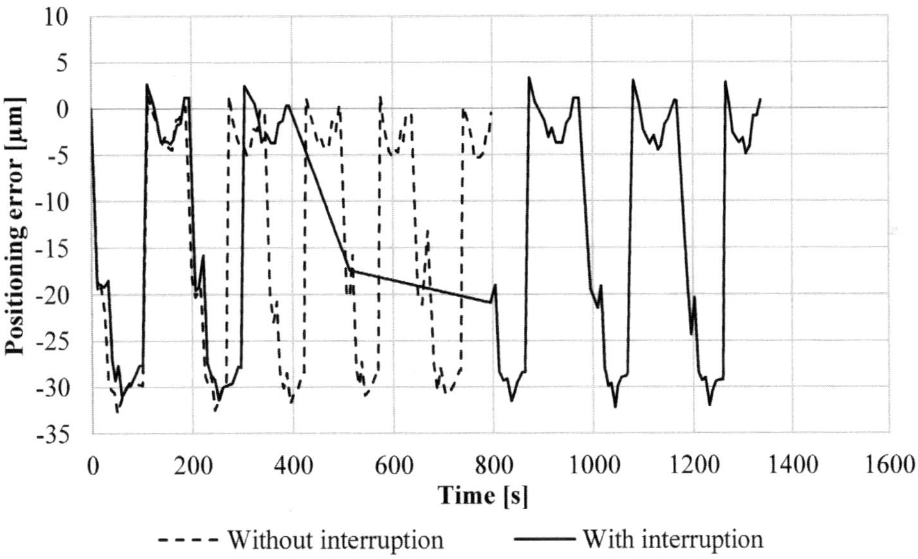

Figure 3. Time dependent results of two interferometric measurements on a CNC machine

Analysis of the effect of the imposed interruption is conducted by comparing the fifth cycle for the interrupted measurement with the fifth cycle of the unimpeded measurement. They are extracted in figure 4 in position and positioning error coordinates. For this machine, the 0 mm position is the forefront extremity, towards the beam splitter, while the 400 mm position is the background extremity of Y the axis travel.

As expected for the CNC machine, important errors are recorded with the interferometer, reaching up to 35 µm. The errors are accurately recorded for both measurement cycles, with a slight difference of approximately 6 µm at the 120 mm position. This might be caused by an increase of clearance in the drive assembly of the machine.

This test shows that a slight interruption, of several minutes, has a low chance of leading to failure of measurements, as the environmental factors do not have a significant variation to affect the results and the slight variations are well taken into account by the environmental compensation unit.

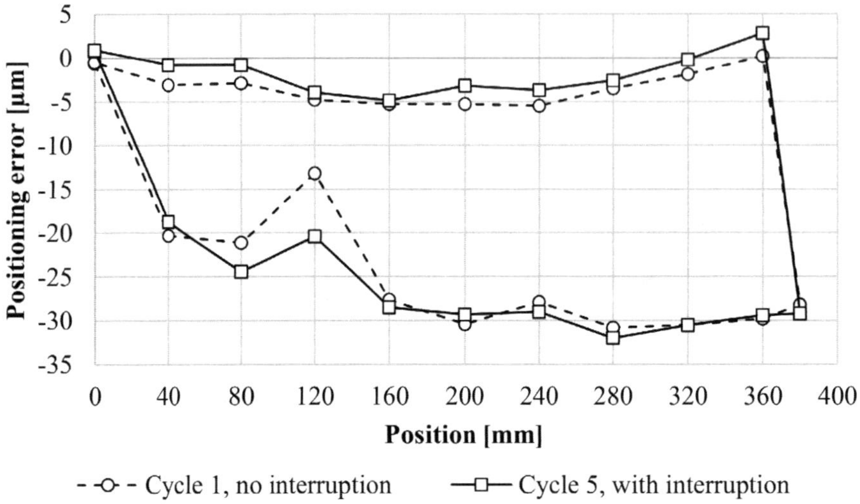

Figure 4. Comparison between the fifth cycles of the measurements

3.3. Increment's implications

ISO 230-2 recommends at least 5 target points (positions) along one meter of axis, for axes up to two meters. Experimental results show that in some cases, a smaller increment can detect punctual differences.

In order to prove this statement, two measuring tests are to be conducted on the same SPINNER MC 1020 CNC machine. The two measuring tests are conducted consecutively, with 20 mm and 40 mm increment respectively for the target positions, as shown in figure 5.

The graphic shows the error value on each measured point. The cycle starts at 0 mm with 0 µm error and has a sequential error increase on the negative scale up to the 280 mm position point from where it continues with an almost constant error of -30 µm up to the 380 mm target position. The return path forces a machine guide's realignment that changes the error from -30 µm to 0 µm and it stays at minimal error for all the returning target points with slight changes but not more than 5 µm, as it can be seen in the graphic.

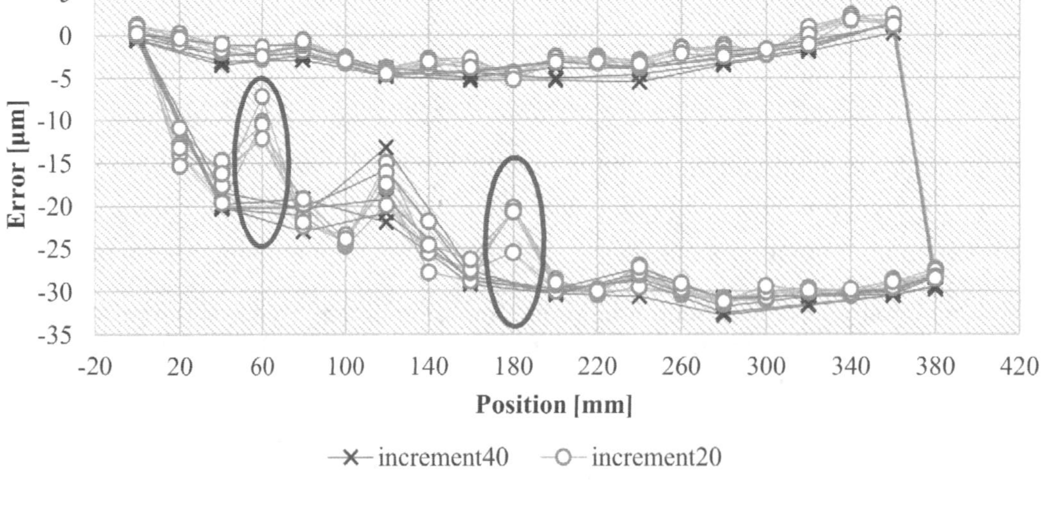

Figure 5. Two interferometric measurements on a CNC machine, with different increments

The increment's implications can be seen at the 60 mm and 180 mm target positions where a repeated deviation of around 10 μm can be observed between the two measurements with different increments. These, along with the one recorded at the 120 mm mark might allude to a possible frequent stop of the spindle at those particular marks, leading to a localised higher wear of the position encoder.

4. Conclusions

The interferometer's applications in the metrological field enhances the measuring accuracy. The measurement procedure involving it is usually detailed by the manufacturer and its usage in positioning accuracy and repeatability evaluation is elaborated by ISO 230-2.

The uncertainties of an ISO 230-2 measurement performed by means of an interferometer are given by various factors. It has been determined that the auto compensation unit is of a great importance, the environmental conditions having a significant impact over the measuring results, due to their influence on the laser wavelength. On the other hand, when the auto compensator is activated and the environmental conditions are steady, a 5-minute pause during the measuring procedure does not influence the measuring result, leading to allow for possible unavoidable interruptions. However, longer periods of time, in the duration of tens of minutes or hours are still considered to alter the results, as atmospheric variations might occur. Another factor is the target positions distance which is to be taken into consideration given that the experimental results showed outcome differences even with increments as small as 20 mm and 40 mm.

When referring to the tested machining centres, the Makino D200Z vertical machining centre has proven to have verry little positioning error, as was expected due to the little to no wear. However, the SPINNER MC 1020 CNC machine has shown in all measurements significant positioning errors that are most likely due to wear that has led to an increase in clearance.

5. References

[1] Bunch B and Hellemans A 2004 *The History of Science and Technology* Boston New York Houghton Mifflin Company

[2] Riehle F, Gill P, Arias F and Robertsson L 2018 The CIPM list of recommended frequency standard values: Guidelines and procedures *Metrologia* **55** 188-200

[3] Vala A, Mewada H, Patel A, Nagora U and Pathak S 2020 Phase detection system based on digital signal processing in millimeter wave interferometer for fusion plasma diagnostics, *Fusion Engineering and Design* **157** 111637

[4] Jang Y S, Wang G, Hyun S et al 2016 Comb-referenced laser distance interferometer for industrial nanotechnology *Scientific Reports* **6** 31770

[5] Shirinzadeh B, Teoh P L, Tian Y, Dalvand M M, Zhong Y and Liaw H C 2010 Laser interferometry-based guidance methodology for high precision positioning of mechanisms and robots *Robotics and Computer-Integrated Manufacturing* **26** 74-82

[6] Artimon (Stochioiu) F P G, Stochioiu C and Popan G 2021 Thermal influence on positioning error and position repeatability of a machining center axes *U.P.B. Scientific Bulletin* **83** **181**

[7] Kawamura S, Ando M, Nakamura T, Tsubono K, Tanaka T, Funaki I et all 2019 The Japanese space gravitational wave antenna - DECIGO *Journal of Physics: Conference Series* **122** 032004

[8] Zihua Z, Chengwei L and Zhen H 2019 Vibration measurement based on multiple Hilbert transform for self-mixing interferometry *Optics Communications* **436** 192-196

[9] Maowei J, Xiaojing L, Shunchuan W, Yongtao G and Linlin G 2011 Use of SAR interferometry for monitoring illegal mining activities: A case study at Xishimen Iron Ore Mine, *Mining Science and Technology* **21** 781–786

[10] Sarychikhina O, Gomez Palacios D, Delgado Argote L A and Gonzales Ortega A 2020 Application of satellite SAR interferometry for the detection and monitoring of landslides along the Tijuana - Ensenada Scenic Highway, Baja California *Journal of South American Earth Sciences* **107** 103030

[11] Iftimia N V, Hammer D X, Bigelow C E, Rosen D I, Ustun T et all 2006 Toward noninvasive measurement of blood hematocrit using spectral domain low coherence interferometry and retinal tracking *Optics Express* **14** 3377-3388

[12] Nissim N, Dudaie M, Barnea I and Shaked NT 2020 Real-Time Stain-Free Classification of Cancer Cells and Blood Cells Using Interferometric Phase Microscopy and Machine Learning *Cytometry A* **99** 511-523

[13] Liu C H, Jiwe W Y, Hsu C C and Hsu T H 2005 Development of a laser-based high-precision six-degrees-of-freedom motion errors measuring system for linear stage *Review of Scientific Instruments* **76** 055110-0551106

[14] Pan F, Nie L, Bai Y, Wang X and Wu X 2017 Geometric errors measurement for coordinate measuring machines *IOP Conference Series: Earth and Environmental Science* **81** 012117

[15] Chen J R, Ho B L, Lee H W, Pan S P and Hsieh T H 2018 Research on Geometric Errors Measurement of Machine Tools Using Auto-Tracking Laser Interferometer *World Journal of Engineering and Technology* **6** 631-636

[16] ISO 230-2:2014 *Test code for machine tools — Part 2: Determination of accuracy and repeatability of positioning of numerically controlled axes* 4th Edition.

[17] Chapman M A V 2013 Environmental compensation of linear laser interferometer readings *Renishaw Technical White Paper TE329* United Kingdom

Acknowledgements

This work was supported by the project "Increasing INCDMTM performance and international involvement by supporting expertise in advanced mechatronic technologies - PERFORM-MECH", financed by The Ministry of Research, Innovation, and Digitalization, in the framework of Programme 1 – Development of the national research and development system, Sub-programme 1.2 – Institutional Performance – Projects for financing excellence in Research, Development and Innovation, Contract no. 3PFE/2021. And a part of the experiments was performed on the equipment of the Research Centre for Intelligent Mechatronic Systems Used for Securing Objectives and Intervention – CERMISO

IOP Conf. Series: Materials Science and Engineering 1268 (2022) 012010 doi:10.1088/1757-899X/1268/1/012010

Analysis and development on general structure and characteristics of laser interferometry systems

M Gheorghe[1], F P G Stochioiu[2,3], D Manolache[1], M R Dijmărescu[1] and D Iliescu[4]
[1] University POLITEHNICA of Bucharest, Manufacturing Engineering Department, Splaiul Independentei, 313, 060042 Bucharest, Romania
[2] PhD Student, University POLITEHNICA of Bucharest, IER Doctoral School, Splaiul Independentei, 313, 060042 Bucharest, Romania
[3] Nat'l Inst. of Res. & Dev. in Mechatronics and Measurement Tech. (INCDMTM), Soseaua Pantelimon, 6 – 8, Sector 2, Bucharest, Romania
[4] Nuclear NDT Research and Services, Soseaua Berceni, 104, Bucharest, Romania

E-mail: marian.gheorghe@upb.ro, flavia.stochioiu@gmail.com, daniel.manolache@upb.ro, manuela.dijmarescu@upb.ro, dragos.iliescu@nuclearndt.ro

Abstract. The laser interferometry systems have been developed for scientific studies, industrial operations, artworks investigation, etc. They present common generic elements, such as laser, optics, beam, polarization, splitting, interferometry, etc., but also diverse specific components and features,such as He-Ne or femtosecond laser, quantum cascade lasers, non-polarising beam splitter, collimating lens, high reflecting mirror, photodetector, attosecond pulse train, etc., as the case. The paper presents, also, a development on general structure and characteristics of laser interferometry systems. Analytical descriptors are introduced for the main implied entities, as well as some qualitative features referring to measurement object, coordinate system, lasers, laser beams, optics, output beams, detectors, environmental sensors and compensation modules, electronics, process software and computer, assembly, calibration, etc. The conclusions accord attention to further laser interferometry development regarding the interacting processes, analytical modelling, simulation, operational characteristics, etc.

1. Analysis on laser interferometry system

The structure and characteristics of a laser interferometry system depend on the task, work type to be performed – research or applied operation, resources, etc.

In order to successively combine a number of laser beams into an output beam, an experimental system was structured mainly as one He-Ne laser and optics, so that, the laser was with two orthogonally longitudinal modes, and the optics consisting in: a collimating lens, a polarizer – for selecting a single longitudinal mode, a half-wave plate (HWP) – to equalize the intensity of beams split, and a polarization beam splitter (PBS), followed by two component combining units - each including a splitter, a high reflecting mirror (HR), a PBS, a quarter-wave plate, a HWP, as well a final part formed by a HR and a PBS, with a high quality output beam [1].

A patented interferometer, for distance measurement with respect to a defined target, is structured as: a laser diode, that emits a laser light beam (LB); optics; detectors (D1, 2, 3). The optics, that processes LB, consists of polarizers (P1, 2), quarter-wave plates (Q1, 2, 3), a half-wave plate (H), polarising beam splitters (PB1, 2), a non-polarising beam splitter (NPB), plane mirrors (PM 1, 2) and a convergent lens (CL). The emitted LB is attenuated by the combination of P1 and PB1, passes through Q1, and is split by NPB1, so that, part of the attenuated beam LB is sent to D1 that records its intensity; the remained part of LB passes through H, enters PB2 that splits it into two orthogonally plane-polarised components B1 and B2, etc.; D2 and D3 measure the fringes produced by superposition of resolved components B1 and B2 [2].

IOP Conf. Series: Materials Science and Engineering 1268 (2022) 012010 doi:10.1088/1757-899X/1268/1/012010

A heterodyne interferometer, with balanced geometrical paths of probe beam (PB) and reference beam (RB), high resistance to environmental influence, usable for measuring refractive indexes in solutions, etc., has been developed as a homemade/ dual frequency/ dual polarization He-Ne laser, optics and high-speed photodiodes (PD1, 2). The optics, that processes B1, 2, consists of polarizers (P1, 2), polarising beam splitters (PBS1, 2, 3), half-wave plates (HWP1, 2), a right angle prism (RP) and a quarter-wave plate. The laser emits two beams (B1, 2), each with two orthogonally main polarization modes. The main polarization modes of B1 are mixed together by P1 and detected by PD1; the main polarization modes of B2 are split into two paths by PBS1, so that, one beam of υ1 frequency is further transmitted as PB, and the reflected beam of υ2 frequency as RB, etc. Finally, the PB and RB are mixed by P2 and transmitted to PD2, for detecting their phase difference [3].

In order to analyse mixing characteristics of laser emission dynamics, an experimental setup is developed as: an external-cavity semiconductor laser unit, one collimation lens (L), two beam splitters (BS1, 2), two neutral density filters (NDF 1, 2), optical reflectors (OR1, 2), one photodetector (PD), and other complementary elements. The laser sends its beam to L and BS1, 2, which separate the beam into a part for a power meter, a part for PD, and transmits the remained part to one of the two units/ cavities, each of these consisting of a NDF and an OR with different attenuation function, i.e., with variable feedback intensity to the laser [4].

A femtosecond laser system is developed, as an erbium-doped fiber laser (EDFL), two lenses, two curved mirrors, a high reflecting mirror, as well as a Cr:ZnS sample, with well-defined characteristics of these components. The system emits pulses with duration of femtoseconds order [5].

To measure geometric characteristics of a transparent plate front and rear surfaces, a laser interferometry system was developed as: a laser diode with an external cavity; optics, consisting in angular tunable diffraction grating (DG), collimation lenses (L1, 2) and beam splitter (BS); recording CCD camera. The laser beam is sent to DG, L1, BS, a reference surface (RS), two surfaces of the target plate, and is reflected back to BS and to the CCD camera, that recorded specific interferograms. By tuning the inclination of DG, the operational laser wavelength was modified in the spectral range of 632.67 - 632.97 nm, during the recording time of 2 min. Thus, the interferograms revealed a plate thickness of 10 ± 0.001 mm, the parallelism of the plate surfaces within 1.2′, for the distance of 145 mm between the RS and the plate front surface [6].

The quantum cascade lasers (QCLs), laser feedback interferometry (LFI)/ self-mixing (SM) interferometry/ effect, as well combination between QCLs and LFI were intensively studied and applied in different systems [7 - 10]. Thus, for instance, within a system based on three-mirror model (M1, 2, 3) of LFI, the laser is represented as an internal cavity, bordered by M1 and M2; the beam leaves the laser through the partially transmissive M2, traverses an external cavity bordered by M2 and M3; a part of this beam re-enters the laser through M2, and mixes with the field from the laser cavity, so that, influences the fundamental laser parameters - emitted power, lasing spectrum, etc. [9].

The interferometry, as a fundamental process, is widely applied in advanced studies. Novel attosecond interferometry systems are developed for high precision measurements. Such of systems include: a free-electron laser, a diffracting split-and-delay unit, a focusing spherical mirror etc. [11], or an infrared beam source, gas cells, an aluminium foil as filter, a curved two-segment mirror and its associated motorized optical iris diaphragm, an attosecond pulse train (APT) target, an extreme ultraviolet (XUV) spectrograph – that resolves the interference of the APT beams, etc. [12].

Process and system elements on laser interferometry for industrial geometric control applications are available in a large diversity. A series of industrial configurations, as well as interferometry and operating data are presented for linear, angular, straightness, etc. measurements [13, 14, 15, etc.].

Thus a regular setup for linear distance measurement is operational in a structure as: a laser source (LS) and a detector (D), incorporated in a laser head; optics, composed by a polarizing beam-splitter (PBS) and retro-reflectors (R1, 2); environmental sensors and compensator (ESC). A laser beam emitted by LS, with a nominal wavelength of 0.633 μm, is split, by PBS, into two beams: one beam is reflected toward the fixed R1, one beam is transmitted to the movable R2, both are reflected again to PBS, and coaxially directed to form an interference beam at D; simultaneously, ESC acts to compensate the eventual wavelength variations [13].

A basic laser measurement system consists of: a He-Ne laser head (L); an interferometer (I) and its associated retroreflector (RR); a measurement receiver (MR); measurement and control electronics. Each of L, I, RR, MR is presented as being relatively insensitive to possible small linear and angular perturbations [14].

The designers should take into consideration that all laser-interferometer positioning systems use a laser head, optics (33% beam splitter or 50% beam splitter, beam bender, linear interferometer, etc.), and electronics, as well as the fact that each of these has certain specifications [15].

2. General structure and characteristics of laser interferometry systems

The present proposed structure and characteristics of laser interferometry systems are developed at high level of generality open to particularizations, in context, as well to other description levels.

2.1. Measurement object

Let measurement object be a set, G, of geometric entities, $G_i, i = \overline{1,n}$,

$$G = \{G_i \mid i = \overline{1,n}\} \Longleftrightarrow G = G_1 \cup G_2 \cup \ldots \cup G_i \cup \ldots \cup G_n \quad (1)$$

It is to underline that each geometric entity G_i is well-defined by the specific features - nominal characteristic, tolerance field, reference elements/ coordinate systems.

The laser interferometry system, LIS, for measurements upon the geometric group G, is structured so that to include elements as coordinate system, laser sources, laser beams, optical modules, output beams, detectors, environmental sensors and compensation modules, process software and computers, etc. as presented in Fig. 1 and subsequently.

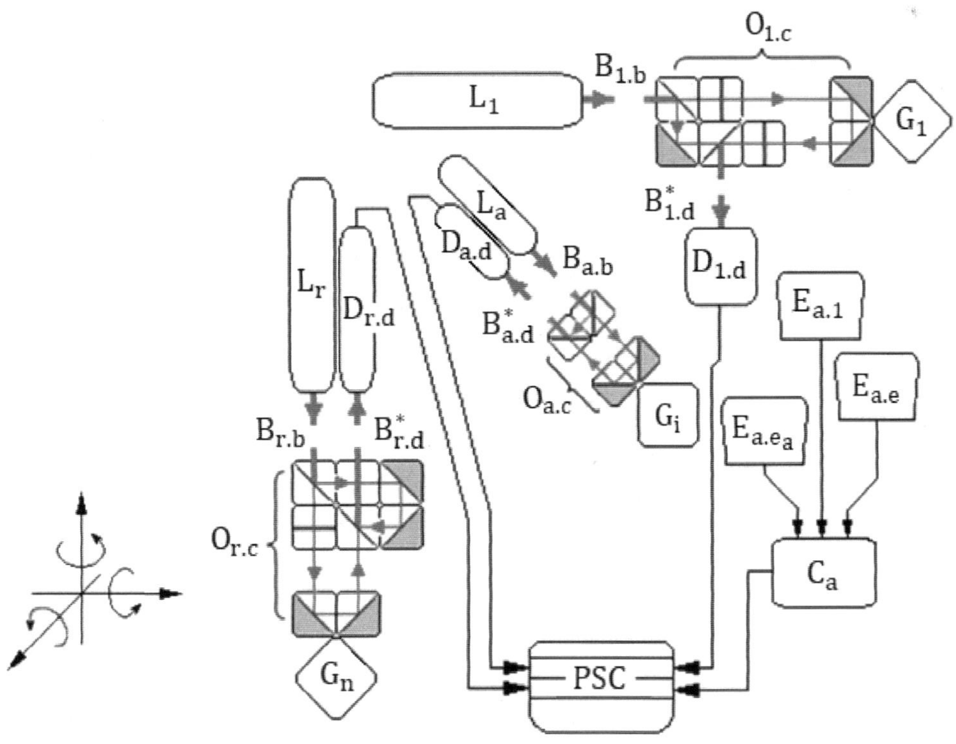

Fig. 1. A general structure of a laser interferometry system

2.2. Coordinate systems

In connection with the reference elements/ coordinate systems associated to the objective G, within LIS, a set, CS, of coordinate systems, CS_j, $j = \overline{1,m}$, is defined/ understood,

$$CS = \{CS_j \mid j = \overline{1,m}\} \Longleftrightarrow CS = CS_1 \cup CS_2 \cup \ldots \cup CS_j \cup \ldots \cup CS_m \quad (2)$$

2.3. Laser sources

The main specific part of LIS consists in a set, L, of laser sources, L_a, $a = \overline{1,r}$,

$$L = \{L_a \mid a = \overline{1,r}\} \Longleftrightarrow L = L_1 \cup L_2 \cup \ldots \cup L_a \cup \ldots \cup L_r \quad (3)$$

Each laser source L_a, $a = \overline{1,r}$, is characterized by dimensions, mass, measurement capability, laser beam properties, ambient conditions, etc.

2.4. Laser beams

Within LIS, each laser source, L_a, $a = \overline{1,r}$, emits one laser beam or a group of laser beams, B_a, $a = \overline{1,r}$, according to the case, that could be expressed as:

$$L_a \Rightarrow B_a, \ a = \overline{1,r} \quad (4)$$

Each laser beam B_a, $a = \overline{1,r}$, exposes specific characteristics, such as size (diameter), divergence, wavelength (color), polarization/ direction, phase, brightness, etc., as well conditions on environment influence.

On the level of LIS, the set, B, of all laser beams is:

$$B = \{B_a \mid a = \overline{1,r}\} \Longleftrightarrow B = B_1 \cup B_2 \cup \ldots \cup B_a \cup \ldots \cup B_r \quad (5)$$

It should be stressed that, among B_a, there is one of the following states:

- each B_a is type of (one) laser beam,
- each B_a is type of group of laser beams,
- some B_a, in number of r_{lb}, are laser beams and the other B_a, in number of r_{glb}, - groups of laser beams, $r_{lb} + r_{glb} = r$.

A group, B_a, $a = \overline{1,r}$, of laser beams, $B_{a.b}$, $a = \overline{1,r}$, $b = \overline{1,b_a}$, is:

$$B_a = \{B_{a.b} \mid a = \overline{1,r},\ b = \overline{1,b_a}\} \Longleftrightarrow B_a = B_{a.1} \cup B_{a.2} \cup \ldots \cup B_{a.b} \cup \ldots \cup B_{a.b_a},\ a = \overline{1,r} \quad (6)$$

In certain cases, among $B_{a.b}$, some laser beams expose tight similar characteristics - as coaxiality, parallelism, frequency, etc., according to the case. So, these can be considered as an unit, $B_{a.b'}$, of laser beams, $B_{a.b'\beta}$, i.e.:

$$B_{a.b'} = B_{a.b} \text{ for } a = \overline{1,r},\ b = \overline{1,b_a},\ b = b',\ b' \in \{1,\ 2, \ldots, b'_a\},\ b'_a \leq b_a \quad (7)$$

and

$$B_{a.b'} = \{B_{a.b.\beta} \mid a = \overline{1,r}, b = \overline{1,b_a}, \beta = \overline{1,\beta_b}\},\ b = b',\ b' \in \{1,\ 2, \ldots, b'_a\},\ b'_a \leq b_a \Longleftrightarrow$$

$$B_{a.b'} = \{B_{a.b'.\beta} \mid a = \overline{1,r}, b' = \overline{1,b'_a}, \beta = \overline{1,\beta_{b'}}\},\ b' \in \{1,\ 2, \ldots, b'_a\},\ b'_a \leq b_a \Longleftrightarrow$$

$$\Longleftrightarrow B_{a.b'} = B_{a.b'.1} \cup B_{a.b'.2} \cup \ldots \cup B_{a.b'.\beta} \cup \ldots \cup B_{a.b'_a.\beta_{b'}},\ a = \overline{1,r},\ b' \in \{1,\ 2, \ldots, b'_a\},\ b'_a \leq b_a \quad (8)$$

If a laser source L_a emits one laser beam, this could be considered as a particular case in the structure described by eq. (6), for $b_a = 1$, i.e.:

$$B_a = \{B_{a.b} | a = \overline{1,r},\ b = \overline{1,b_a},\ b_a = 1\} <=> B_a = B_{a.1}, a = \overline{1,r} \quad (9)$$

or in the structure described by eq. (8), for $b'_a = 1$ and $\beta_{b'} = 1$, i.e.:

$$B_{a.b'} = \{B_{a.b'.\beta} | a = \overline{1,r}, b' = \overline{1,b'_a}, \beta = \overline{1,\beta_{b'}}\}, b'_a = 1,\ \beta_{b'} = 1 <=> B_{a.1} = B_{a.1.1}, a = \overline{1,r} \quad (10)$$

From eqs. (9) and (10), it results:

$$B_a = B_{a.1} = B_{a.1.1}, a = \overline{1,r} \quad (11)$$

Eq. (11) shows that when a laser source L_a emits only one laser beam, this is B_a, $B_{a.1}$ or $B_{a.1.1}$, $a = \overline{1,r}$, as the case.

It is to be noted that a (individual/ elementary) laser beam is: type of B_a, $a = \overline{1,r}$, or $B_{a.b}$, $a = \overline{1,r}$, $b = \overline{1,b_a}$, if B_a or $B_{a.b}$ is clearly characterized as laser beam (not as group or unit of laser beams); any of $B_{a.b'\beta}$, $a = \overline{1,r}$, $b' = \overline{1,b'_a}$, $\beta = \overline{1,\beta_{b'}}$.

Also, each laser beam is a monochromatic, directional and coherent beam.

Example

A laser L_1 emits a group B_1 of laser beams, that consists of one laser beam $B_{1.1}$ and an unit $B_{1.2}$ of two laser beams, $B_{1.2.1}$ and $B_{1.2.2}$. Thus: $B_1 = B_{1.1} \cup B_{1.2}$ and $B_{1.2} = B_{1.2.1} \cup B_{1.2.2}$, equivalent to $B_1 = B_{1.1} \cup B_{1.2.1} \cup B_{1.2.2}$; each of the laser beams $B_{1.1}$, $B_{1.2.1}$ and $B_{1.2.2}$ is a monochromatic, directional and coherent beam.

2.5. Optical modules

Within LIS, a laser beam or a group of laser beams B_a, $a = \overline{1,r}$, is transmitted toward an optical module or a group of optical modules, O_a, $a = \overline{1,r}$, according to the case, that could be expressed as:

$$B_a \Rightarrow O_a,\ a = \overline{1,r} \quad (12)$$

Each optical module or group of optical modules O_a, $a = \overline{1,r}$, is defined by overall characteristics as dimensions, mass, relative position, etc.

On the level of LIS, the set, O, of all optical modules is:

$$O = \{O_a | a = \overline{1,r}\} <=> O = O_1 \cup O_2 \cup \ldots \cup O_a \cup \ldots \cup O_r \quad (13)$$

It should be emphasized that, among O_a, there is one of the following states:

- each O_a is type of optical module,
- each O_a is type of group of optical modules,
- some O_a, in number of r_{om}, are optical modules and the other O_a, in number of r_{gom}, - groups of optical modules, $r_{om} + r_{gom} = r$.

A group, O_a, $a = \overline{1,r}$, of optical modules, $O_{a.c}$, $a = \overline{1,r}$, $c = \overline{1,c_a}$, is:

$$O_a = \{O_{a.c} | a = \overline{1,r},\ c = \overline{1,c_a}\} <=> O_a = O_{a.1} \cup O_{a.2} \cup \ldots \cup O_{a.c} \cup \ldots \cup O_{a.c_a}, a = \overline{1,r} \quad (14)$$

An optical module could be considered as a particular case in the structure described by eq. (14), for $c_a = 1$, i.e.:

$$O_a = \{O_{a.c} | a = \overline{1,r},\ c = \overline{1,c_a},\ c_a = 1\} <=> O_a = O_{a.1}, a = \overline{1,r} \quad (15)$$

Thus, an optical module that receives one laser beam B_a, $B_{a.1}$ or $B_{a.1.1}$, $a = \overline{1,r}$, or an unit $B_{a.b'}$, $a = \overline{1,r}$, $b' = \overline{1,b'_a}$, of laser beams $B_{a.b'\beta}$, $a = \overline{1,r}$, $b' = \overline{1,b'_a}$, $\beta = \overline{1,\beta_{b'}}$ [see eqs. (11), (8)], can be O_a or $O_{a.1}$, $a = \overline{1,r}$.

Each optical module $O_{a.c}$ or O_a, $a = \overline{1,r}$, $c = \overline{1,c_a}$, could be considered as a subgroup of optical elements, $O_{a.c.\theta}$, $a = \overline{1,r}$, $c = \overline{1,c_a}$, $\theta = \overline{1,\theta_c}$, or $O_{a.1.\theta}$, $a = \overline{1,r}$, $c = 1, \theta = \overline{1,\theta_c}$, respectively, i.e.:

$$O_{a.c} = \{O_{a.c.\theta} | a = \overline{1,r}, c = \overline{1,c_a}, \ \theta = \overline{1,\theta_c}\} <=>$$

$$<=> O_{a.c} = O_{a.c.1} \cup O_{a.c.2} \cup \ldots O_{a.c.\theta} \cup \ldots \cup O_{a.c.\theta_c}, a = \overline{1,r} \tag{16}$$

or, in the case of O_a, $a = \overline{1,r}$ [see eq. (15)]:

$$O_a = O_{a.1} = \{O_{a.1.\theta} | a = \overline{1,r}, c = 1, \ \theta = \overline{1,\theta_c}\} <=>$$

$$<=> O_a = O_{a.1} = O_{a.1.1} \cup O_{a.1.2} \cup \ldots O_{a.1.\theta} \cup \ldots \cup O_{a.1.\theta_c}, a = \overline{1,r} \tag{17}$$

- Within an optical module, one or more optical elements are under well-defined action(s) associated to a geometric entity G_i, $i = \overline{1,n}$. Let these optical elements be $O_{a.c.\theta'}$ or $O_{a.1.\theta'}$, $a = \overline{1,r}$, $c = \overline{1,c_a}$, $\theta' \in \{1, 2, \ldots, \theta_c\}$, from the optical module $O_{a.c}$, $a = \overline{1,r}$, $c = \overline{1,c_a}$, or from the optical module O_a, $O_a = O_{a.1}$, $a = \overline{1,r}$, respectively, i.e.:

$$O_{a.c.\theta'} \in O_{a.c} = \{O_{a.c.\theta} | a = \overline{1,r}, c = \overline{1,c_a}, \ \theta = \overline{1,\theta_c}\}, \theta = \theta', \ \theta' \in \{1, 2, \ldots, \theta_c\} <=>$$

$$<=> O_{a.c.\theta'} = O_{a.c.1} \text{ or/and } O_{a.c.2} \text{ or/and } \ldots, a = \overline{1,r}, c = \overline{1,c_a} \tag{18}$$

or, in the case of O_a, $O_a = O_{a.1}$, $a = \overline{1,r}$ [see eq. (17)]:

$$O_{a.1.\theta'} \in O_{a.1} = \{O_{a.1.\theta} | a = \overline{1,r}, c = 1, \ \theta = \overline{1,\theta_c}\}, \theta = \theta', \ \theta' \in \{1, 2, \ldots, \theta_c\} <=>$$

$$<=> O_{a.1.\theta'} = O_{a.1.1} \text{ or/and } O_{a.1.2} \text{ or/and } \ldots, a = \overline{1,r} \tag{19}$$

An optical element $O_{a.c.\theta}$, $a = \overline{1,r}$, $c = \overline{1,c_a}$, $\theta = \overline{1,\theta_c}$, reveals a main function – as polarizing beam splitter, beam splitter, dual beam splitter, beam reflector, collimating lens, etc., specific functions - as ratio of reflecting/ transmitting of incident laser beam, bending angle of laser beam, etc., as well constructive characteristics as dimensions, mass, relative position, etc.

2.6. Output beams of optical modules

Within LIS, an optical module or a group of optical modules O_a, $a = \overline{1,r}$, receives an incident laser beam or a group of incident laser beams, B_a, $a = \overline{1,r}$, and processes it through specific actions, such as polarization, splitting, reflection, refraction, interferometry, so that its output is one output beam or a group of output beams, B_a^*, $a = \overline{1,r}$, that could be expressed as:

$$O_a \Rightarrow B_a^*, a = \overline{1,r} \tag{20}$$

Each output beam or group of output beams B_a^*, $a = \overline{1,r}$, has to have defined characteristics type of size, divergence, wavelength, phase, relative position, etc., as well as conditions on environment influence level.

On the level of LIS, the set, B^*, of all output beams is:

$$B^* = \{B_a^* | \ a = \overline{1,r}\} <=> B^* = B_1^* \cup B_2^* \cup \ldots \cup B_a^* \cup \ldots \cup B_r^* \tag{21}$$

It should be stressed that, among B_a^*, there is one of the following states:

- each B_a^* is type of (one) output beam,
- each B_a^* is type of group of output beams,
- some B_a^*, in number of r_{ob}, are output beams, and the other B_a^*, in number of r_{gob}, - groups of output beams, $r_{ob} + r_{gob} = r$.

A group, B_a^*, $a = \overline{1,r}$, of output beams, $B_{a.d}^*$, $a = \overline{1,r}$, $d = \overline{1,d_a}$, is:

$$B_a^* = \{B_{a.d}^* | a = \overline{1,r},\ d = \overline{1,d_a}\} <=> B_a^* = B_{a.1}^* \cup B_{a.2}^* \cup \ldots \cup B_{a.d}^* \cup \ldots \cup B_{a.d_a}^*,\ a = \overline{1,r} \quad (22)$$

In certain cases, among $B_{a.d}^*$, some output beams expose tight similar characteristics - as coaxiality, wavelength, etc. So, these can be considered as an unit, $B_{a.d'}^*$, of output beams, $B_{a.d'.\rho}^*$, i.e.:

$$B_{a.d'}^* = B_{a.d}^* \text{ for } a = \overline{1,r},\ d = \overline{1,d_a},\ d = d',\ d' \in \{1,\ 2, \ldots, d_a'\},\ d_a' \le d_a \quad (23)$$

and

$$B_{a.d'}^* = \{B_{a.d'.\rho}^* | a = \overline{1,r},\ d' = \overline{1,d_a'},\ \rho = \overline{1,\rho_{d'}}\},\ d' \in \{1,\ 2, \ldots, d_a'\},\ d_a' \le d_a <=>$$

$$<=> B_{a.d'}^* = B_{a.d'.1}^* \cup B_{a.d'.2}^* \cup \ldots \cup B_{a.d'.\rho}^* \cup \ldots \cup B_{a.d'.\rho_{d'}}^*,\ a = \overline{1,r},\ d' \in \{1,\ 2, \ldots, d_a'\},\ d_a' \le d_a \quad (24)$$

An important *unit of output beams* is consisting in *coaxial output beams* which, in certain well-defined conditions, *interfere* and form an *interference beam.*

An output beam could be considered as a particular case in the structure described by eq. (22), for $d_a = 1$, i.e.:

$$B_a^* = \{B_{a.d}^* | a = \overline{1,r},\ d = \overline{1,d_a},\ d_a = 1\} <=> B_a^* = B_{a.1}^*,\ a = \overline{1,r} \quad (25)$$

or in the structure described by eq. (24), for $d_a' = 1$ and $\rho_{d'} = 1$, i.e.:

$$B_{a.d'}^* = \{B_{a.d'.\rho}^* | a = \overline{1,r}, d' = \overline{1,d_a'}, \rho = \overline{1,\rho_{d'}}\}, d_a' = 1, \rho_{d'} = 1 <=> B_{a.1}^* = B_{a.1.1}^*,\ a = \overline{1,r} \quad (26)$$

From eqs. (25) and (26), it results:

$$B_a^* = B_{a.1}^* = B_{a.1.1}^*,\ a = \overline{1,r} \quad (27)$$

Eq. (27) shows that when an optical module O_a generates only one output beam, this is B_a^*, $B_{a.1}^*$ or $B_{a.1.1}^*$, $a = \overline{1,r}$, as the case may be.

It should be noted that an (individual/ elementary) output beam is: type of B_a^*, $a = \overline{1,r}$, or $B_{a.d}^*$, $a = \overline{1,r}$, $d = \overline{1,d_a}$, if B_a^*or $B_{a.d}^*$ is clear characterized as output beam (not as group or unit of output beams); any of $B_{a.d'.\rho}^*$, $a = \overline{1,r}$, $d' = \overline{1,d_a'}$, $\beta = \overline{1,\rho_{d'}}$.

2.7. Detectors of output beams

Within LIS, an output beam or a group of output beams B_a^*, $a = \overline{1,r}$, is transmitted toward a detector (receiver) or group of detectors, D_a, $a = \overline{1,r}$, as the case, that could be expressed as:

$$B_a^* \Rightarrow D_a,\ a = \overline{1,r} \quad (28)$$

Each detector or group of detectors, D_a, $a = \overline{1,r}$, is defined by overall characteristics as dimensions, mass, relative position, sensitivity, heat dissipation, output signal, etc.

On the level of LIS, the set, D, of all detectors is:

$$D = \{D_a |\, a = \overline{1,r}\} \Longleftrightarrow D = D_1 \cup D_2 \cup \ldots \cup D_a \cup \ldots \cup D_r \quad (29)$$

It should be emphasized that, among D_a, there is one of the following states:

- each D_a is type of detector,
- each D_a is type of group of detectors,
- some D_a, in number of r_d, are detectors and the other D_a, in number of r_{gd}, - groups of detectors, $r_d + r_{gd} = r$.

A group, D_a, $a = \overline{1,r}$, of detectors, $D_{a.d}$, $a = \overline{1,r}$, $d = \overline{1,d_a}$, is:

$$D_a = \{D_{a.d} | a = \overline{1,r},\ d = \overline{1,d_a}\} \Longleftrightarrow D_a = D_{a.1} \cup D_{a.2} \cup \ldots \cup D_{a.d} \cup \ldots \cup D_{a.d_a},\ a = \overline{1,r} \quad (30)$$

One detector could be considered as a particular case in the structure described by eq. (30), for $d_a = 1$, i.e.:

$$D_a = \{D_{a.d} | a = \overline{1,r},\ d = \overline{1,d_a},\ d_a = 1\} \Longleftrightarrow D_a = D_{a.1},\ a = \overline{1,r} \quad (31)$$

If $D_a = D_{a.1}$, $a = \overline{1,r}$ [see eq. (31)], then only one output beam B^*_a, $B^*_{a.1}$ or $B^*_{a.1.1}$, $a = \overline{1,r}$, or an unit $B^*_{a.d'}$, $a = \overline{1,r}$, $d' = \overline{1,d'_a}$, of output beams $B^*_{a.d'.\rho}$, $a = \overline{1,r}$, $d' = \overline{1,d'_a}$, $\rho = \overline{1,\rho_{b'}}$ [see eqs. (27), (24)], is transmitted to the detector D_a (or $D_{a.1}$), $a = \overline{1,r}$, as the case.

2.8. Environmental sensors and compensation modules

Within LIS, a group, E_a, $a = \overline{1,r}$, of environmental sensors, E_{ae}, $a = \overline{1,r}$, $e = \overline{1,e_a}$, register and transmit specific information to a compensation module, C_a, $a = \overline{1,r}$,

$$E_a = \{E_{a.e} | a = \overline{1,r},\ e = \overline{1,e_a}\} \Longleftrightarrow E_a = E_{a.1} \cup E_{a.2} \cup \ldots \cup E_{a.e} \cup \ldots \cup E_{a.e_a},\ a = \overline{1,r} \quad (32)$$

and

$$E_a \overset{\cdots}{\rightarrow} C_a,\ a = \overline{1,r} \quad (33)$$

On the level of LIS, the set, E, of all groups of environmental sensors E_a, $a = \overline{1,r}$, is:

$$E = \{E_a |\, a = \overline{1,r}\} \Longleftrightarrow E = E_1 \cup E_2 \cup \ldots \cup E_a \cup \ldots \cup E_r \quad (34)$$

and the set, C, of all compensation modules C_a, $a = \overline{1,r}$, is:

$$C = \{C_a |\, a = \overline{1,r}\} \Longleftrightarrow C = C_1 \cup C_2 \cup \ldots \cup C_a \cup \ldots \cup C_r \quad (35)$$

The environmental sensors and the compensation module are defined by specific characteristics, such as dimension, mass, measured and processed environmental parameter – air temperature, air pressure, air humidity, material temperature, vibration, etc., range, accuracy, interface, and so forth.

2.9. Electronics modules

Within LIS, a series of electronics modules, for processing the specific signals, are associated to certain system parts among laser sources, optics, detectors, compensation units, as appropriate.

2.10. Process software and computer(s)

The LIS structure includes process software and computer(s), PSC, for the specific data acquisition, analysis, recorder, etc.

2.11. Assembly, calibration and other elements

The LIS structure includes a series of specific elements as tripods, magnetic bases, adjustable mounts, fixing screws, cables, shutters, etc. for positioning and fixing of components system, data transmitting, calibration, protection, etc.

2.12. LIS real structure and characteristics

The real constructive structure and characteristics of LIS depend of the measurement object features and other techno-economic conditions.

Thus, a *laser source* element and a *detector* (*receiver*) element are distinct system components or - incorporated together in a *laser unit* component, optical elements type *beam splitter* and *interferometer* are distinct or - incorporated together in a *beam splitter*, some *environmental sensors* are included in a *compensation module*, etc.

Remark

The above general structure and characteristics of laser interferometry systems, which have been introduced through a selection of analytical descriptors and qualitative features, can be used as a reference base for: further presentations on detailed scientific and operational characteristics of measurement objects, lasers, laser beams, optical elements etc., e.g., by building-up specific databases; configuring of future experimental set-ups, and determination of their optimal components, in order to satisfy the implied measurement conditions; continuous research on interacting processes, analytical modelling, etc.

3. Conclusions

The structure and characteristics of laser interferometry systems are designed and implemented function of the work type – research or applied operation, and its attributes, resources, etc.

The laser interferometry systems present common generic elements, such as laser, optics, beam, etc., but also specific components and features, such as He-Ne or femtosecond or free-electron laser, non-polarising beam splitter, attosecond pulse train, etc., as the case. A series of system configurations and features for industrial geometric control applications are available for linear, angular, etc. measurements.

The development on general structure and characteristics of laser interferometry systems includes analytical descriptors and some general features referring to measurement object, coordinate system, lasers, laser beams, optics, output beams, detectors, environmental sensors and compensation modules, electronics, process software and computer(s), assembly, calibration and other elements.

Further laser interferometry developments should envisage the interacting processes, analytical modelling, simulation, operational characteristics, etc.

4. References

[1] Dong H C, Li X, Wei C Y, He H B, Zhao Y N and Shao J D 2009 Coaxial combination of coherent laser beams, *Chin. Opt. Lett.* **7** (11) 1012-1044

[2] Speake C C, Aston S M, Pena Arellano F E and Copland T P E 2010 Interferometer, Patent No. US 2010/0238456 A1

[3] Yoon S, Park Y and Cho K 2013 A new balanced-path heterodyne I/Q – interferometer scheme for low environmental noise, high sensitivity phase measurements for both reflection and transmission *Opt. Express* **21** (18) 20722-20729

[4] Onea C, Sterian P E, Andrei I R and Pascu M L 2019 High frequency chaotic dynamics in a semiconductor laser with double-reflector selective cavity *UPB Sci. Bull. Series A* **81** (4) 261-270

[5] Vasilyev S, Moskalev I, Smolski V, Peppers J, Mirov M, Fedorov V, Martyshkin D, Mirov S and Gapontsev V 2019 Octave-spanning Cr:ZnS femtosecond laser with intrinsic nonlinear interferometry *Optica* **6** (2) 126-127

[6] Sun T, Zheng W, Yu Y, Asundi A K and Valyukh S 2019 Determination of surface profiles of transparent plates by means of laser interferometry with wavelength tuning *Opt. Lasers Eng.* **115** 59-66

[7] Vitiello M S, Scalari G, Williams B and Natale P 2015 Quantum cascade lasers: 20 years of challenges *Opt. Express* **23** (4) 5167-51182

[8] Lim Y L, Bertling K, Taimre T, Gillespie T, Glenn C, Robinson A, Indjin D, Han Y, Li L, Linfield E H, Davies A G, Dean P and Rakić A D 2019 Coherent imaging using laser feedback interferometry with pulsed-mode terahertz quantum cascade lasers *Opt. Express* **27** (4) 10221-10233

[9] Rakić A D, Taimre T, Bertling K, Lim Y L, Dean P, Valavanis A and Indijin D 2019 Sensing and imaging using laser feedback interferometry with quantum cascade lasers *Appl. Phys. Rev.* **6** 021320

[10] Qi X, Agnew G, Taimre T, Han S, Lim Y L, Bertling K, Demić A, Dean P, Indjin D and Rakić A D 2020 Laser feedback interferometry in multi-mode terahertz quantum cascade lasers *Opt. Express* **28** (10) 14246-14262

[11] Usenk S, Przystawik A, Jakob MA, Lazzarino L L, Brenner G, Toleikis S, Haunhorst C, Kip D and Laarmann T 2017 Attosecond interferometry with self-amplified spontaneous emission of a free-electron laser *Nat. Commun.* **8** 15626

[12] Azoury D, Kneller O, Rozen S, Bruner B D, Clergerie A, Mairesse Y, Fabre B, Pons B, Dudovich N and Krüger M 2019 Electronic wavefunctions probed by all-optical attosecond interferometry *Nature Photon.* **13** (2) 54-59

[13] ***, Renishaw, XL-80 laser system training course manual/ Part 1, Part 2, Part 3, Brochure: XL- 80 laser measurement system, etc. www.renishaw.com (accessed on 16.08.2022)

[14] ***, Agilent Laser and Optics, User's Manual, Volume I, https://www.uzimex.cz/soubory/20080403_laser_optika_cast_1.pdf (accessed on 19.08.2022)

[15] ***, Keysight Technologies, Optics and Laser Heads for Laser-Interferometer Positioning Systems, https://www.keysight.com/us/en/assets/7018-06711/technical-overviews/5964-6190.pdf (accessed on 10.09.2022)

IOP Conf. Series: Materials Science and Engineering 1268 (2022) 012011 doi:10.1088/1757-899X/1268/1/012011

Testing Mechatronic Systems and their Integration into the Digital Enterprise

L M Sima

PhD student, University POLITEHNICA of Bucharest, Faculty of Industrial and Robotics Engineering, Department of Robots and Production Systems, Bucharest, Romania

e-mail: liviusima23@gmail.com

Abstract. When a mechatronic system is built, certain considerations prevail. The product needs to follow certain steps in production. Here is discussed the need to test an industrial mechatronic system during production and the interfaces of its components. As mechatronic systems tend to become increasingly complex, the test methods should include certain strategies that allow the delivered to the customer with the desired specifications and in a timely manner. With the help of the digital factory, a product is stored electronically and a new or an improved variant of the product could be at any time created, with less effort in production, before this one becomes uninteresting on the market.

1. Introduction
The mechatronic system contains elements of type as mechanical, electrical and electronic. [1] These form a whole, controlled by a software program, loaded into a controller. As technology has evolved, there have been demands for systems to perform increasingly complex functions. Testing at an early stage is essential to remove any errors in the system. [2]
A digital factory is a relatively new concept and offers many advantages in diagnosing the occurrence of such errors in production, as well as a more precise follow-up of the construction. This can be completely digital, thus being able to test the interfaces of the mechatronic system at any time and even being able to simulate products superior to the one desired by the customer. Such an improved product is later used, especially after analyzing the first version of the delivered product. The testing of the interfaces is done manually in the digital factory, but also automatically, following the desired models from the requirements of the mechatronic system. [3]

2. New hardware and software test models
As well as existing testing approaches, there are testing methods in terms of abstraction and generation: [4]
- based on models, where design testing, case testing and script generation are used;
- engine / test execution platforms, where the script is run and the result is reported;
- Hardware interface and configuration with automatic testing infrastructure (ATI), where hardware and monitoring interfaces (sensors, actuators), hardware in the loop (HiL), software in the loop (SiL) are tested.

ATI supports testing systems in different implementations with configurations and scenarios. One such approach focuses on delivering software interfaces. Such an approach focuses to provide software interfaces with devices on the hardware devices. The test scripts are allowed to run on different hardware configurations. Thus, a tester manually generates scripts and executes them on the system configuration. In particular, this automated testing infrastructure supports strategies such as "loop hardware" and "loop simulator (or software)", allowing mechatronic systems to be in a variety of tested configurations. Model-based testing are approaches used in test generation, based on an abstract case test with specifications of a system written on a high-level in specification languages, standardized such as UML. The changing occurs on the existing algorithms, which are combined with various testing strategies. The idea is to cover the system requirements through testing. [5] Unified Modeling Language (UML) is a modelling language, standardized, by which developers can specify, visualize, build, and document the artifacts of a software system. Thus, UML creates scalable, secure and robust artifacts in execution. UML is involved in the software development of object-oriented programs. Regarding the design methodology related with mechatronics, the basic content is that of the improved "V" model. [6] The advantage of an improved model is that of the independence of the way the mechatronic product is manufactured. The model created can be applied in the classic projects managed, but also in projects which are using the agile methodology. [7]

The proposal of a "V" model with software development engineering processes occurred appeared in 1995 and it was developed by Bröhl and Dröschel. [8] The "V" test pattern is the breakdown of the system into its components (left side of the “V”) and the gradual integration of elements and subsystems into the entire technical system (right side of the “V”). [9] Furthermore, the characteristics of the developing product are verified and validated. [10] A properly created system is developed in terms of "validation" requirements in the desired "verification" manner. [11] In the 2004 version of VDI 2206, the necessities are graphically represented as an entry area. The system implementation is displayed at the bottom of the "V" model. Mechatronic systems engineering involves several disciplines, such as mechanical, software, electrical/electronic, optical, hydraulic, pneumatic or other. The "V" model leads to its desired result. Between the two parts of the "V" pattern, the arrows depict the provision of properties that include the review and approval activities. Title insurance is listed from right to left, giving the impression of a retroactive process. [12]

Systems engineering consists of a structured, multidisciplinary engineering approach to mechanical systems that aims for an interdisciplinary optimum in a given period and funds. To this end, the disciplines are organized and connected to each other using certain patterns. In addition to the interdisciplinary optimum, the basic qualities of systems engineering often stated are protection and credibility, complexity management, customer satisfaction, reduced delivery period and reduced costs. The emphasis falls on interdisciplinary and frontloading that highlight the focus of management and the spending of large capacities in the initial stages of the development procedure. Systems engineering totalizes managerial organizational and technical tasks that are brought together collectively. As well as the "V" model in the 2004 VDI 2206 guide, there are a large number of "V" models that are similarly structured as the VDI guide. These variants have been focused on individual applications and specific areas. An improved model represents the entire life cycle in Figure 1; others focus only on the product development stage. [13] [13] One method to consider is that the representation and description of modeling and analysis in today's "V" model should be universal so that the prominence of modeling in improvement is properly perceived. Too much variety of "V" patterns can cause confusion and, furthermore, incorrect application of the steps to be followed. The new testing model proposed consists of three layers:

- basic activities and tasks (collection of information on requirements, system architecture and design, implementation of system elements, system integration and verification, validation and transition to the next phase);
- handling and working with requirements (requirements management);

- modeling and analysis activities (specifications, architecture, implementation, integration).

The three layers graphically contain the disciplines involved at the base of the "V" pattern. The graph symbolizes the execution of system elements based on profound and interconnected details between the disciplines involved. These disciplines are important to the core component as well as requirements engineering modeling and analysis. A concise explanation of the significant factors of this testing model is provided. [14] This proposed testing model is adapted from existing models from software testing methodologies and can be used for mechatronic systems.

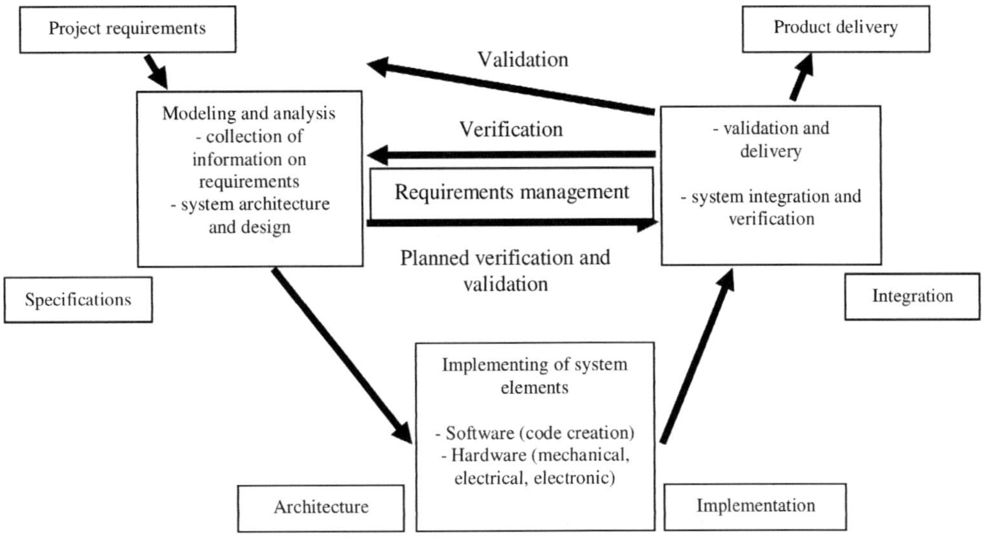

Figure 1. Testing model for a mechatronic system

The product development process consists of modeling, structuring and analyzing the requirements. Regarding "stakeholders", a list or a diagram of requirements is derived and documented in a technical format and in a very formal way. Having initially stated their requirements, which are not described in the first stage of the system, their necessities and desires are later transformed into system demands.

Specifications (result of requirements modeling) are a description of the developed system with a large number of demands. [15] This happens very often. In order to develop as complete a specification as possible, a list of key features is created for the mechatronic system. This list helps the user to get the completeness and good quality of the specifications. The fundamental characteristics for a product specification are completeness, certainty, consistency and precision. [16]

During the development of the product, the list of requirements is completed by the successive concretization and detailing of the product's qualities. The goal is for all essential requirements to be quantitatively defined by the time the system architecture is complete. Therefore, requirements may change due to feedback and iterations from downstream sections or by updating client requests or market demands. Changes are therefore systematically supervised and assessed. The output of requirements modeling and analysis is a description of the system's features to be met. Target parameters to ensure system properties are derived from requirements. In terms of system architecture and design, the focus of interdisciplinary product engineering occurs in the system architecture and design phase. [17]

In the architecture of the system, a structure of global interdisciplinary solution, is developed for the system to be created. In accordance with the practical configuration and the fundamental premises of action (below), the system planning includes the mechanical structure of the product, signal flow structures and electronic circuit representations, as well as the structuring of a software program into its modules and components, including those interfaces from a software perspective. Thus, the system architecture embodies elements, relationships and principles necessary for the design and further development of a system. There may be alternatives in terms of structure, compared based on comprehensible criteria. These would be interdependencies, functional analysis and consideration of exclusion criteria. This type of evaluation is carried out with the participation of users. The best possible task of specific functions is determined by an iterative procedure based on system partitioning. The job of the system architect is to break down the system into implementable units. Starting from the functions and properties required by the system, a system is systematically broken down in terms of functionality, properties and implementation. The output is quantified into verifiable units such as system components with specific requirements assigned and relevant interface information. The number and scheme of interfaces within an architecture and system limit considerably influence the straightforwardness, flexibility and testability of a system. When implementing system elements, they are dimensioned, designed and detailed.

Implementation consists of conversion an algorithm or software design into a computer program. Having defined the software architecture, the functions are inserted using algorithms, and then the related code is created. Database models are implemented by providing modeling in concrete schemes. In addition, the user or system interface is designed. Mechanical, electronic and software interfaces are considered in detail when implemented. At the same time, the implementation of the elements of the system serves to solve problems that arise during the inclusion of system components. [18]

In terms of system integration and verification, integration gradually combines the elements and subsystems implemented by the system into the superior level of the product hierarchy. The goal is for the integrated system to meet the requirements specified above. The correlation of the constituents or elements of the disciplines involved is thoroughly reviewed in the form of a system imitation or virtual empowering. The inclusion strategy must be agreed upon in advance with the important stakeholders. Property forecasting and insurance are done during the development period. The properties of the system to be created are anticipated at an initial phase based on the models. Property validation accompanying development is used for constant inspection of the aspired solution based on investigations with virtual/physical prototypes (computer simulation / hardware experiment) or their "hybrid" (HiL, for example). [19]

Loop hardware (HiL) indicates the techniques of checking (testing) electronic control units (e.g. machine control units) by connecting them to special banks that more or less completely reproduce the electrical system, and electronic for which they are intended. These banks are complex devices that reproduce in whole or in part, physically and / or software, the product for which the test units are intended. These units consist of the following components:

- sensors, i.e. input devices (some of those available on cars are the buttons to control the window lifts and other mechanisms, to measure the oil temperature, the sound wheels for ABS, etc.);
- the actuating devices, i.e. the devices that perform an action when controlling the control unit (the engine that actually raises and lowers the windows of a car, the pump that controls the brake caliper, etc.);
- electrical and electronic cables and relative connections (i.e. the electrical structure and the network that interconnects the control units with the relative terminations where to physically connect sensors, actuators and control units);
- models, i.e. computer programs that emulate what is not physically part of the bank by "cheating" the actually connected devices (for example, the engine with all its sensors and actuators).

Due to the close interdependence between integration and ownership, it is necessary to plan at which phases the system will be incorporated and for which areas of application it is needed, using which

IOP Conf. Series: Materials Science and Engineering 1268 (2022) 012011 doi:10.1088/1757-899X/1268/1/012011

method, and when and at what level of application the implementation will be proven. Thus, the following approaches to property insurance are differentiated:

- theoretical examination: calculation, modeling and imitation;
- checkup: for easily noticeable, measurable and acknowledged properties;
- demonstration: high-quality proof of utility, usually with basic instrumentation;
- test: quantitative evidence, usually in an established test environment; inclusion in the superior level of the system must not begin until the inspection of the first level has been finished.

Hardware simulation in the loop (HiL) or HWIL is a method used in the improvement and testing of complex real-time embedded systems. It provides an efficient program by adding the intricacy of the product under control to the test platform which is later on included in testing and development by adding a mathematical representation of all related dynamic systems. These mathematical representations are also known as "factory simulation". The embedded system to be tested interacts with this factory simulation. HiL simulation should include electrical emulation of sensors and actuators. These electrical emulations act as an interface between installation simulation and embedded system testing. The value of each electrically emulated sensor is controlled by the installation simulation and is read by the embedded system under test (feedback). The tested embedded system also implements its control algorithms by issuing actuator control signals. Changes in the control signals lead to changes in the variable values in the plant simulation. Similar to HiL, there is a software simulation in the loop (SiL), which refers to the software components of the mechatronic system. Unlike looping hardware, the looping software (SiL) method does not need any special hardware. The original code in the ECU is written to a file, usually a dynamic link library (DLL), using an automated encoding process. This DLL can be integrated into an environment model in a simulation tool (for example, Matlab Simulink). The DLL is then tested in the world of purely virtual models. Unlike the HiL method, no control devices and no real-time simulators are required. SiL simulators do not run in real time and the simulation can be stopped at any time for troubleshooting. Hardware running problems or variable overruns that occur on the real ECU (electronic control unit) can hardly be detected with SiL simulations. An advantage of SiL is that the target hardware has not yet been determined and the costs are much lower due to the lack of a simulation environment. The simulation model used here can also be used in subsequent HiL simulations and thus individual tests can be compared with each other.

In particular, strategies such as HiL ("loop hardware") and SiL ("loop simulator") are supported, allowing built-in (incomplete) testing and mechatronic systems in various configurations. While environments such as these address diversity, external dependencies, internal setup, and the sustenance of test infrastructure, their deployment, configuration, and usage are mostly manual, laborious and open to error.

Each pattern and document required for subsequent stages of the product life cycle is stored and structured using a Product Data Management (PDM) system to manage design data and engineering processes in one location throughout the development process of the product. Then, the end product certification is built and finalized. [20]

As a prerequisite for a later digitally identical copy, essential models are created to enable simulation of the system's behavior, also named Digital Master. The digital depiction of related client-specific product versions is called a Digital Twin. This is an updated representation of an actual physical asset in operation. [21] As a description of a current asset, a digital twin indicates the current condition of the asset and contains important historical information about the property. "Digital Twins" is later used to anticipate subsequent behavior, improve control or maximize performance. A "digital twin" can model a unit, a structure of components, or a system of structures, such as power plants, pumps, motors.

The methods of using techniques such as brainstorming, interviews, surveys/ questionnaires, prototyping and document analysis and requirements analysis generate computer-assisted models. Consequently, a substantial number of models are generated during system architecture development and implementation. As shown in the testing model, Figure 1, modeling and examination also impact particular disciplines: each subject involved has a diversity of models. In the conditions of model-driven improvement and MBSE (model-based systems engineering), topic models are connected to system-level models and actively exchange information. Model formation also requires vigorous analysis and fusion. It refers to the assessment, impression and simulation of components of the system, subsystem or system as a whole. [22]

To emphasize the importance of requirements management as a basis for a successful project, the terms are used individually in the "V" model. The organizing, allocation, review and inclusion of requirements changes need to be controlled during the development project. Diverse tools, databases and approaches assist requirements handling. [23]

Different levels of the system have a great influence on each other, so changes in demands are reflected and registered for the entire project and for the relevant project participants. The three tiers symbolize the primary tasks of the testing model and are endorsed by a six-point control structure. In the middle of the model, between the two "V" shapes, three arrows symbolically suggest the activities of securing goods. Singular arrows are characteristic of activities that can be highly important at any moment. "Verification and Validation Planning" describes the need to start extensive activities already on the left side of the "V" model to ensure the requirements defined in the sequence. Planning goes beyond writing a good requirement. It describes not only the way a claim should be obtained, but also the way it should be made. [24]

Verification: transferred to technical structures, it is a necessity to understand whether the way something is done (a software program) correlates with the specification (description of algorithms). Verification is generally done formally. Validation: it is necessary to understand if the product is appropriate for the proposed purpose or reaches the wanted value. The anticipations of the technical adept and the user matter at this point. In everyday language, validation is showing that a suitable product is being developed. Thus, it should not be done formally. Within mechatronic systems, the basics of development processes, the approach to interdisciplinary modeling, the importance of iterations for knowledge acquisition, different types of testing and the basics of model-based design are followed. A system can be validated with prototypes or virtual models. [25]

Validation procedures and strategies include:

- ✓ iterative adaptation: this approach is remarkable in that there are a large number of iterations, a continuous adaptation of development according to a changing knowledge base and a sequential development. The development stages are a bit premeditated and the interfaces between the teams exist only in a moderate way, not with detailed specifications. Virtual models are used only moderately.

Advantages and disadvantages: low development risk, because a functional system is always available. But development is slow and intensive resources are needed.

- ✓ focus on subsystem development: this strategy is characterized by a detailed analysis of all components. The development of the subsystems is parallel and the physical validation of the subsystems is carried out in detail. The integration of the whole system and its validation (general system level test) is done at an advanced stage.

Advantages and disadvantages: each subsystem is developed in detail. Late integration is a risk in the sense that the subsystems do not work together properly and large iterations occur.

- ✓ global system validation: this strategy is characterized by a parallel development of the subsystems and early integration into the general system. Validation is mainly performed on the general level of the system.

Advantages and disadvantages: all subsystems are tested very early, generally according to the function implementation system. However, the level of maturity of the subsystems is partially and not high enough, so some subsystems need to be reassembled in order to be optimized.

- ✓ detailed virtual modeling: in this strategy, a top-down approach with integrated SysML modeling can be clearly seen with a distinct definition of interfaces between teams, with clear responsibilities within the team, sophisticated CAD design and advanced use of virtual models. Physical validation of sub functions takes place very late.

Systems Modeling Language (SysML) is a general-purpose modeling language for systems engineering applications. It supports the specification, analysis, design, verification and validation of a wide range of systems and systems-to-systems. Advantages and disadvantages: all subsystems have been virtually tested and may have very rapid developmental progress during an observed preconceived process. However, only a few functional prototypes can be developed, which is why some critical points are identified very late. In general, it can be seen that some of the validation characteristics change during the development of phases. When we consider the system level, there is a major change from the first to the second project stage. In the first phase, it is very important to validate unique requirements or sub-functions at the level of the discipline-specific subsystems. On the contrary, in the second phase the level changes the interdisciplinary subsystems. This is understandable because later in the project, the greater the integration of the individual components, the more complex the subsystems are and a new validation must be performed.

In mechatronic systems, certain technical parameters that are always considered can be highlighted for validation. As shown in the testing model, the mechatronic system is validated according to the needs of the project stakeholders. Parameters to consider are the elements in Figure 1. These technical (quantifiable) parameters are followed in the test so that the mechatronic system is validated.

3. Industrial processes in the digital enterprise

The digital enterprise refers to a platform that has the concept of a software company behind it. This digital platform is purely software and serves as an intermediary between users, consumers and service providers. An electronic platform is the technology behind a digital model. By managing the business relationships within models, a digital business platform enables access to new revenue streams by collaborating and co-creating joint contributions to an ecosystem. For example, a business application (such as booking.com) is a standard provider of a product, a value chain between the company, or consumers (those who want to rent a hotel and the related steps). With the help of the digital platform, digital services are offered to consumers, giving a much better experience to the users, so that an improved product can be delivered. New services exist for end users, making the digital platform useful. A diagram for the digital platform is shown in the next image.

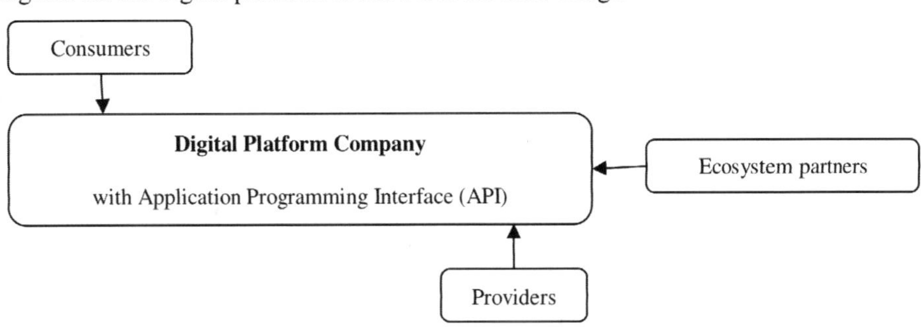

Figure 2. Digital Platform Company

In the industrial world, the notion of digital platform is developing very fast, and this concept is being used on a higher scale. The idea behind it is that every sold product becomes connected, which means that all possible data is saved on the digital platform.

After that, the "digital twin" can be configures, because all the information of the physical system is known on the digital medium and, using all that data plus all data versions, services for users can be built.

We can study the product information, forecast, to develop new services for end users. A diagram for the digital platform in the industry is shown in the following figure.

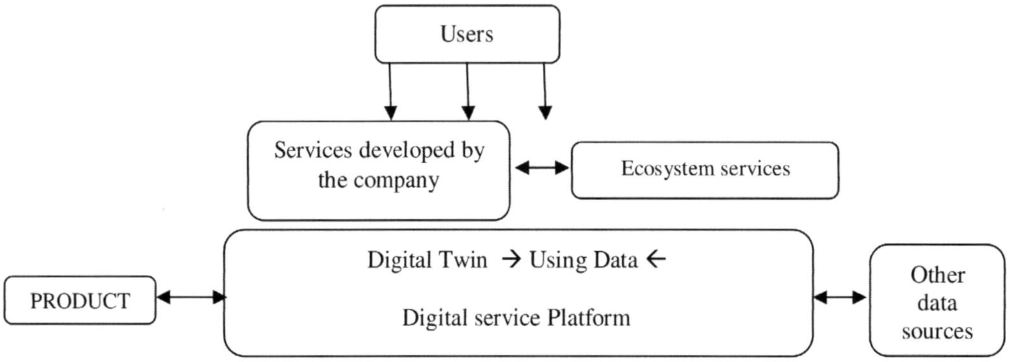

Figure 3. Digital Platform in industry

Ecosystems provide basic services that make human life possible. The core of the digital revolution is capitalizing on data to deliver new services on a digital platform, and this is the kind of revolution that is coming into business.

The digital platform will take on more and more value, and there is a risk that the company (which originally made the product) will be interrupted when another company easily creates a new digital platform or product like the one originally made. That new company will take over the value chain and the product will become more common.

When a company does not have the advantage of a digital platform, it cannot have the possibility to produce an already made product, with better qualities in the future. At least that is the way it is, but on a much smaller scale. Without this strategic advantage, the factory virtually undermines its production for a better quality. This creates a strategic advantage over its competitors. Nowadays, a company should not give up making such an investment and increase its production, because it puts a competing company on the line to take its place.

A company needs to be active in the digital platform revolution, always ready and actively working on those digital services. In terms of technologies, there are 4 areas in core digital technologies:

- connectivity - all products are connected; data can be sent in terms of usage, monitoring, connections. In addition, data and users can connect to the web services provided;
- Big Data - to extract the value from all created data;
- artificial intelligence (AI) - can be used in the product itself, in the decision support for the product to become autonomous; automation of tasks previously performed by the operator that can be automated by AI;
- cyber security - no data loss, it is mandatory to secure all data to make sure that it cannot be altered or hacked.

References

[1] Sima LM (2022) *Using Arduino PLC and Robot within Mechatronic Systems*, International Journal of Ad hoc, Sensor & Ubiquitous Computing (IJASUC), February 2022, Volume 13, Number 1, Chennai, India, ISSN (print) 0976-2205, (online) 0976-1764;

[2] Sima LM, Zapciu M (2020) *Connections and Interfaces of Mechatronic Components on Digital Factory*, Conference Proceedings of the Academy of Romanian Scientists, PRODUCTICA Scientific Session, Volume 12, Number 1, May 2020, Bucharest, Romania, ISSN (print) 2067-2160, (online) 2067-9564, pg. 129-140;

[3] Sima LM, Zapciu M (2020) *The nonconventional mechatronic system*, Nonconventional Technologies Review, Volume 24, Number 2, June 2020, Sibiu, Romania, ISSN (print) 2359-8646, (online) 2359-8654, pg. 4-13;

[4] Silva Filho RS, Budnik CJ, *An Integrated Model-Driven Approach for Mechatronic Systems Testing*, ISBN 978-1-4577-1906-6 IEEE, Montreal, Canada, 2012;

[5] Sima LM, Zapciu M (2022) *Testing the Interfaces for Mechatronic Systems*, Journal of Mechatronics and Robotics, Vol. 6, Issue 1, April 2022, Dubai, UAE, ISSN (print) 2617-0345, (online) 2617-0353, pg. 22-27;

[6] Sima LM, Zapciu M (2021) *A Testing Model for the Mechatronic System*, U.P.B. Sci. Bull., Series D, Vol. 83, Issue 1, January 2021, Bucharest, Romania, ISSN (print) 1454-2358, (online) 2286-3699, pg. 113-124;

[7] Bathelt J, Jonsson A, Bacs C, et al, *Applying the new VDI Design Guideline 2206 on mechatronic systems controlled by a PLC*, International conference on engineering design (ICED 05), Melbourne, Australia, 15-18 August 2005, ISBN 0-85825-788-2, pg. 2601-2614;

[8] Bröhl AP, Dröschel W, *Das V-Modell: Der Standard für die Softwareentwicklung mit Praxisleitfaden*, Oldenbourg Wissenschaftsverlag, 2. Aufl. Edition, 1995, ISBN 978-3486234701, 656p;

[9] Graham D, Black R, van Veenendaal E, *Foundations of Software Testing: ISTQB Certification*, Cengage Learning, Stamford, USA, 2019, ISBN 978-1-4737-6479-8, Ch. 2,5;

[10] Gräßler I, Hentze J, *The new V-Model of VDI 2206 and its validation* at - Automatisierungstechnik, vol. 68, no. 5, pp. 312-324, 2020;

[11] Spillner A, Linz T, Schaefer H, *Software Testing Foundations: A Study Guide for the Certified Tester Exam*, 4th Edition, Rocky Nook, California, USA, 2014, ISBN 978-1-937538-42-2, Ch. 3-6;

[12] Gausemeier J, Moehringer S, *New Guideline VDI 2206 - A flexible procedure model for the design of mechatronic systems*, Proceedings of ICED 03, the 14th International Conference on Engineering Design, Stockholm Sweden, August 2003, ISBN 1-904670-00-8, pg. 1-10;

[13] Sima LM, Zapciu M (2020) *Production Methodology of the Mechatronic System*, Proceedings of the 36th International Business Information Management Association (IBIMA), ISBN 978-0-9998551-5-7, 4-5 November 2020, Granada, Spain, published in Norristown, USA, pg. 7032-7039;

[14] Sima LM (2022) *Design and Testing Ways for Mechatronic Systems*, International Journal on Cybernetics & Informatics (IJCI), Vol. 11, No. 1/2, April 2022, Chennai, India, ISSN (print) 2320-8430, (online) 2277-548X, pg. 61-69;

[15] Walden DD, Roedler GJ, Forsberg K, Hamelin RD, Shortell TM, *Systems engineering handbook: A guide for system life cycle processes and activities*, 4th ed. Wiley, New Jersey, USA, 2015, ISBN 978-1-118-99940-0, Ch. 2,9;

[16] Gräßler I, Haas V, Suchowerskyj W, *Innovation Based on Applying Design Methodology* in TMCE 2012, pg. 37–43;

[17] Dolga V, *Proiectarea sistemelor mecatronice*, Ed. Politehnica, 2007, ISBN 978-973-625-573-1, pg. 1-60;

[18] Dolga V, *Mecatronică. Teoria sistemelor*, Ed. Politehnica, 2011, ISBN 978-606-554-235-8;

[19] Daigl M, Glunz R, *ISO 29119 – Die Softwaretest-Normen verstehen und anwenden*, dpunkt.verlag, Heidelberg, Germany, 2016, ISBN 978-3-86491-772-1, Teil II;

- state-of-the-art technology - the digital platform is built on state-of-the-art technology, with open APIs and an architecture of native containerized micro-services in the cloud;
- fast implementation and configuration - the ability to get up and running in just a few weeks and dodge avoid long and intricate implementation programs;
- quick and easy integration - extended storage of industry standard open APIs for quick and easy incorporation of current infrastructure;
- scalable and flexible – made with vigorous and versatile business functions, intended for protection, expandability and high availability;
- information is used in the business ecosystem - full visibility is obtained on all points of view of the business and operations, along with ecosystem offers, using analysis and business information;
- innovative monetization capabilities - enabled by a state-of-the-art billing engine, adaptable catalog and full partner agreement aptitudes accessible to all enterprise partners in the ecosystem;
- strong portfolio of support services - a strong portfolio of support services to ensure success - including fast delivery.

Regarding the mechatronic system tested, the industrial interfaces are checked with the created software programs, and the parameters of the components are found. After testing, it is decided whether the component meets the system demands and the production of the mechatronic structure is continued or not. The advantage is that this is done digitally, and all processed information is saved in the cloud. For physical testing of mechatronic elements, this can be done in terms of EMC, ESD and vibration resistance. In this case, the types of international certifications must be followed.

4. Trends in the model testing

Automated testing has become one of the leading technologies for software development (Dev) and IT operations (Ops) or DevOps. Companies invest time and effort to offer end-to-end delivery opportunities. Because of these desires to deliver faster, end users have become virtually testers of the product. A company needs to ensure that quality is not compromised due to "fast" delivery. Testing is one of the most important DevOps checks that a company can use to ensure that those customers can provide useful data about the delivered product. This includes access control, activity logging, traceability and disaster recovery. Slow feedback cycles, slow development loops and developer productivity will remain priorities. Quality and access control are preventive, while others are reactive. In the future, there will be an increasing emphasis on quality, as it prevents customers from having a bad experience. Thus, the rapid delivery of value - or rather, the rapid delivery of the right value at the right level of quality - is the key trend in the future. Some trends to follow are:

✓ test automation: test automation efforts will continue to accelerate. A large number of companies still perform manual tests before delivering the product. This is a big disadvantage, because it cannot be delivered quickly if there are many people involved, thus slowing down the delivery process. An exception is exploratory testing, in which humans are a necessity. Automating manual testing is a time-consuming process. Many organizations have at least some form of testing automation. Therefore, this will remain one of the top trends in the future.

✓ continuous maintenance of quality: as teams automate testing and adopt DevOps, quality must become part of DevOps. This means that we will have the quality of a common responsibility of everyone in the organization. The company's experts change the tests to create new "catch" error streams. Thus, it is known that the validation is done earlier than it should be (according to the "V" model), in order to speed up the delivery. Teams must prioritize tests according to the desired goal (requirements). We need to change the tests early to detect problem. Changing large test suites is difficult because it does not have the opposite effect of causing long delays while running tests in an anterior part of the workflow (according to the "V" model). Many companies label some tests in a large suite to run before joining, but the downside is that these tests may or may not be relevant to a

IOP Conf. Series: Materials Science and Engineering 1268 (2022) 012011 doi:10.1088/1757-899X/1268/1/012011

particular set of changes. The selection of predictive tests provides a good solution for running relevant tests only.

✓ data-based DevOps: in recent years, the industry has focused on connecting different tools by building robust delivery directions. Each of these tools produces a great amount of data, but that data is used to a minimum or not at all. It has gone from "artisanal" solutions to the "scale" stage in the evolution of tools in the delivery directions. The next step is to bring intelligence to the tools. Emphasis will be placed on data-driven decision-making.

✓ artificial intelligence (AI): test generation tools: there are at least two major test problems: insufficient testing and too many tests. Test-generating tools can solve related to insufficient testing. Either to create a UI (user-interface) test, we have to write a lot of code or a tester has to manually click on the UI, which is a slow process. To avoid wasting time by involving the tester, test generation tools use artificial intelligence to create and run UI (human-machine interface tests) on different platforms. For example, a tool once explored by a team allows actions to be recorded in a web application to create tests without scripts. Although testing without scripts is not a new idea, what is new is that this tool tests in step with the changes made to the user interface. Another exploration tool is related to AI robots that act like humans. They touch buttons, scroll through images, type text, and navigate screens to detect problems. Once they find a problem, they report a bug in an application (tool like Jira, bug tracking, etc.) for the developers to act. It is believed that more testing tools that use artificial intelligence will gain ground in the near future. This would be another direction of research.

✓ a big and real problem: the execution time of the test: automatic testing is more used, but many teams are struggling to make the transition from manual testing to automatic testing. Continuous quality culture is becoming part of DevOps. The tools will continue to become smarter. Testing tools will help reduce the gap between manual and automatic testing. As teams add more tests, they face real problems with *test run time*. Although several tests help to improve quality, they often become an obstacle to productivity. Automatic running will help in the near future.
These are the future directions regarding the testing of mechatronic systems interfaces from the digital enterprise.

5. Conclusions

In the article, the importance of testing mechatronic systems in the production process of the digital enterprise is highlighted. Once the system is defined as requirements, it must be "broken down" into parts (mechanical, electrical and electronic), which are then integrated and (subsequently) assembled as a whole. A choice of a manufacturing method based on a particular model must be carefully considered, based on system information and previous experience of the team.

The proposed testing model, with the three layers presented, ensures better coordination in the production process of a mechatronic system. A relatively new concept is the digital factory (enterprise), which incorporates in software all the elements of a mechatronic system, and a new product can be made with less effort. In addition to the possibility that an already delivered product may be re-created at any time, there is the possibility that another similar but improved product may be delivered.

Everything that is newly implemented is tested automatically, with few mistakes, because the whole production process is well documented. Everything can be done remotely, from the "cloud", through interconnected teams.

A digital factory can deliver better quality mechatronic systems from one version to another. As technology has evolved, there has been a tendency to deliver a product faster. The new testing model is a way to realize a mechatronic system, where the manufacturing steps are not lost or cannot be forgotten.

References

[1] Sima LM (2022) *Using Arduino PLC and Robot within Mechatronic Systems*, International Journal of Ad hoc, Sensor & Ubiquitous Computing (IJASUC), February 2022, Volume 13, Number 1, Chennai, India, ISSN (print) 0976-2205, (online) 0976-1764;

[2] Sima LM, Zapciu M (2020) *Connections and Interfaces of Mechatronic Components on Digital Factory*, Conference Proceedings of the Academy of Romanian Scientists, PRODUCTICA Scientific Session, Volume 12, Number 1, May 2020, Bucharest, Romania, ISSN (print) 2067-2160, (online) 2067-9564, pg. 129-140;

[3] Sima LM, Zapciu M (2020) *The nonconventional mechatronic system*, Nonconventional Technologies Review, Volume 24, Number 2, June 2020, Sibiu, Romania, ISSN (print) 2359-8646, (online) 2359-8654, pg. 4-13;

[4] Silva Filho RS, Budnik CJ, *An Integrated Model-Driven Approach for Mechatronic Systems Testing*, ISBN 978-1-4577-1906-6 IEEE, Montreal, Canada, 2012;

[5] Sima LM, Zapciu M (2022) *Testing the Interfaces for Mechatronic Systems*, Journal of Mechatronics and Robotics, Vol. 6, Issue 1, April 2022, Dubai, UAE, ISSN (print) 2617-0345, (online) 2617-0353, pg. 22-27;

[6] Sima LM, Zapciu M (2021) *A Testing Model for the Mechatronic System*, U.P.B. Sci. Bull., Series D, Vol. 83, Issue 1, January 2021, Bucharest, Romania, ISSN (print) 1454-2358, (online) 2286-3699, pg. 113-124;

[7] Bathelt J, Jonsson A, Bacs C, et al, *Applying the new VDI Design Guideline 2206 on mechatronic systems controlled by a PLC*, International conference on engineering design (ICED 05), Melbourne, Australia, 15-18 August 2005, ISBN 0-85825-788-2, pg. 2601-2614;

[8] Bröhl AP, Dröschel W, *Das V-Modell: Der Standard für die Softwareentwicklung mit Praxisleitfaden*, Oldenbourg Wissenschaftsverlag, 2. Aufl. Edition, 1995, ISBN 978-3486234701, 656p;

[9] Graham D, Black R, van Veenendaal E, *Foundations of Software Testing: ISTQB Certification*, Cengage Learning, Stamford, USA, 2019, ISBN 978-1-4737-6479-8, Ch. 2,5;

[10] Gräßler I, Hentze J, *The new V-Model of VDI 2206 and its validation* at - Automatisierungstechnik, vol. 68, no. 5, pp. 312-324, 2020;

[11] Spillner A, Linz T, Schaefer H, *Software Testing Foundations: A Study Guide for the Certified Tester Exam*, 4th Edition, Rocky Nook, California, USA, 2014, ISBN 978-1-937538-42-2, Ch. 3-6;

[12] Gausemeier J, Moehringer S, *New Guideline VDI 2206 - A flexible procedure model for the design of mechatronic systems*, Proceedings of ICED 03, the 14th International Conference on Engineering Design, Stockholm Sweden, August 2003, ISBN 1-904670-00-8, pg. 1-10;

[13] Sima LM, Zapciu M (2020) *Production Methodology of the Mechatronic System*, Proceedings of the 36th International Business Information Management Association (IBIMA), ISBN 978-0-9998551-5-7, 4-5 November 2020, Granada, Spain, published in Norristown, USA, pg. 7032-7039;

[14] Sima LM (2022) *Design and Testing Ways for Mechatronic Systems*, International Journal on Cybernetics & Informatics (IJCI), Vol. 11, No. 1/2, April 2022, Chennai, India, ISSN (print) 2320-8430, (online) 2277-548X, pg. 61-69;

[15] Walden DD, Roedler GJ, Forsberg K, Hamelin RD, Shortell TM, *Systems engineering handbook: A guide for system life cycle processes and activities*, 4th ed. Wiley, New Jersey, USA, 2015, ISBN 978-1-118-99940-0, Ch. 2,9;

[16] Gräßler I, Haas V, Suchowerskyj W, *Innovation Based on Applying Design Methodology* in TMCE 2012, pg. 37–43;

[17] Dolga V, *Proiectarea sistemelor mecatronice*, Ed. Politehnica, 2007, ISBN 978-973-625-573-1, pg. 1-60;

[18] Dolga V, *Mecatronică. Teoria sistemelor*, Ed. Politehnica, 2011, ISBN 978-606-554-235-8;

[19] Daigl M, Glunz R, *ISO 29119 – Die Softwaretest-Normen verstehen und anwenden*, dpunkt.verlag, Heidelberg, Germany, 2016, ISBN 978-3-86491-772-1, Teil II;

[20] ISO/IEC/IEEE 29119-1:2022, *Ingénierie du logiciel et des systèmes — Essais du logiciel — Partie 1*: Concepts et définitions, Geneva, Switzerland;

[21] Schäuffele J, Zurawka T, *Automotive Software Engineering: Grundlagen, Prozesse, Methoden und Werkzeuge effizient einsetzen*, Springer-Verlag, Berlin, Germany, 2010, ISBN 978-3-8348-0364-1;

[22] Stark R, Damerau T, *Digital Twin*, in International Academy for Production Engineering, Chatti S, Laperrière L, Reinhart G, Tolio T (eds), CIRP Encyclopedia of Production Engineering, Springer Berlin Heidelberg, Germany, 2019, ISBN: 978-3-642-35950-7;

[23] Trauer J, Schweigert-Recksiek S, Engel C, Spreitzer K, Zimmermann M, *What is a Digital Twin? – Definitions and Insights from an Industrial Case Study in Technical Product Development*, International Design Conference – Design 2020, Cavtat, Croatia;

[24] Pohl K, Hönninger H, Achatz R, Broy M, *Model-Based Engineering of Embedded Systems: The SPES 2020 Methodology*, Springer Berlin Heidelberg, Germany, 2012, ISBN 978-3-642-34613-2;

[25] Ebert C, *Systematisches Requirements Engineering: Anforderungen ermitteln, spezifizieren, analysieren und verwalten*, 6th ed. dpunkt.verlag, Heidelberg, Germany, 2019, ISBN 978-3-96088-452-1.

IOP Conf. Series: Materials Science and Engineering 1268 (2022) 012012 doi:10.1088/1757-899X/1268/1/012012

Aspects Regarding the Modelling and Design of 3D-printed Bending Soft Pneumatic Actuators

C Rad[1], O Hancu[1], C Lapusan[1]

[1] Department of Mechatronics and Machine Dynamics, Technical University of Cluj-Napoca, 400114 Cluj-Napoca, Romania

E-mail: olimpiu.hancu@mdm.utcluj.ro

Abstract. Soft robotics has the potential to revolutionize our society and industries due to its increased adaptability and soft interaction with environment. In this context, the paper presents the results of finite element modelling and design of two 3D-printed bending soft pneumatic actuators made from NinjaFlex® flexible filament. Geometrical parameters of the actuators were chosen based on the recommendations from the literature and NinjaFlex® material elastic properties were modelled using Ogden hyperelastic model. The simulation was implemented in ANSYS® Workbench software and both designs were compared and analysed in terms of bending angle performance and equivalent (von-Mises) stresses for an input pressure of 0.1÷0.5 MPa with an increment of 0.1 MPa. The obtained simulation results can be used as a design guideline for 3D-printed bending soft pneumatic actuators made from flexible thermoplastic polyurethanes.

1. Introduction

In the last decade, *soft/continuum* robots have seen an exponential growth in the scientific literature compared with the classic rigid robots [1-3]. Soft robots are made from highly flexible materials and structures such as hydrogels, elastomers, thermoplastic polyurethanes (TPUs), electroactive polymers, etc. which allows an increased adaptability and soft interaction with the environment [4].

Actuators represent the foundation of soft robotics and various soft actuation technologies have been proposed and investigated in the literature over the years. Some of them includes tendon driven mechanisms, ionic polymer-metal composites, shape memory alloys, dielectric elastomers, shape memory polymers, fluidic actuators (pneumatic and hydraulic), etc. [5,6]. However, pneumatic actuation using Soft Pneumatic Actuators (SPAs) remains the main technology in the filed as highlighted in [7,8].

SPAs can be categorized in: (i) fiber-reinforced actuators (for example McKibben actuators) and (ii) PneuNets/bellow-type actuators (a network of pneumatic chambers/bellows). Both categories of SPAs can be pre-programmed from the design phase to execute different types of motion when are pressurized/vacuumized. Thus, extension/contraction, bending, twisting/helical, or bidirectional/omnidirectional motion can be achieved by choosing the right geometric parameters of the actuator and the material type [8].

In figure 1 is presented for example a commercial *soft gripper* with two bending SPAs fingers made from silicone that can be used for grasping soft/fragile objects [9]. The bending motion of the fingers in grip mode can be controlled with a pressure signal from 0÷1.2 bar in this case. In release mode the chambers are vacuumized.

a) b) c)

Figure 1. Soft gripper with two bellow-type SPAs fingers [9]: a) fingers relaxed, b) fingers in grip mode c) fingers in release mode

Usually, the bending behavior of SPAs is designed/pre-programmed priori to fabrication to reduce the development time. This implies to understand how geometry of the chambers and material properties is influencing the bending performance of the actuator [10,11]. Such answers can be obtained through modelling. Modeling of SPAs can be accomplished using analytical, statistical (data-driven) and numerical models. Obtaining accurate analytical models is challenging due to nonlinear material properties and large deformations of the actuator under pressurization [12]. Statistical models try to overcome these problems by capturing material, and geometrical features of the actuator through experimental data sets [9] but requires a real prototype of SPA to measure those data. Numerical models are usually based on Finite Element Modelling (FEM) implemented in software packages like ANSYS® or Abaqus® and requires high computational resources and a detailed CAD model of the actuator [13]. From the three modelling approaches presented above, FEM seems to be a good candidate for handling complex geometrical designs, material nonlinearities and large deformation behavior of SPAs in the design stage. At the same time, eliminates the need of a complex analytical model and the existence of a real prototype of the actuator.

Regarding SPAs manufacturing, the most common technique implies the use of molds to cast the actuator and its internal geometry. However, the method is time-consuming, involves many intermediary steps and requires special equipment (vacuum pumps for degassing the mixture, curing ovens, etc.) [7, 11]. More recently, direct 3D printing techniques were successfully used to manufacture SPAs by using direct-drive 3D-printers optimized for printing flexible filaments. Very complex geometries and faster times between SPAs prototypes can be achieved in this way but requires choosing the right set of printing parameters [8,10]. Thus, considering the issues of molding techniques and the general availability of 3D-printers nowadays, the 3D-printing of SPAs becomes an interesting option to be explored in research and development activities that involves soft robotics/actuators.

In this context, the paper proposes an approach for FEM modelling and simulation of 3D-printed bellow-type SPAs made from NinjaFlex® TPU flexible material [14]. The approach considers the main design parameters (geometry and material properties) and their influence on bending behavior of SPAs. The constraints of 3D-printing manufacturing process are considered too. The main contribution of this paper is related to the application of the proposed approach on two SPAs with different geometry.

2. Design parameters considerations

Bending SPAs can generate large bending deformations with one input pressure source. Bending performance is directly influenced by the geometry of the actuator and material type. Therefore, geometrical and material parameters are considered critical design parameters [8,15-17].

The influence of geometrical parameters such as gap size g between two chambers, wall thickness t_1, bottom layer thickness t_2, chamber width w, number of chambers n and the type of cross section (rectangular, half-round and honeycomb designs are best suited for larger bending angles for the same input pressure compared with round designs) on bending angle of SPAs was studied in [15,16,18,19] and authors conclusions are presented synthetically in figure 2.

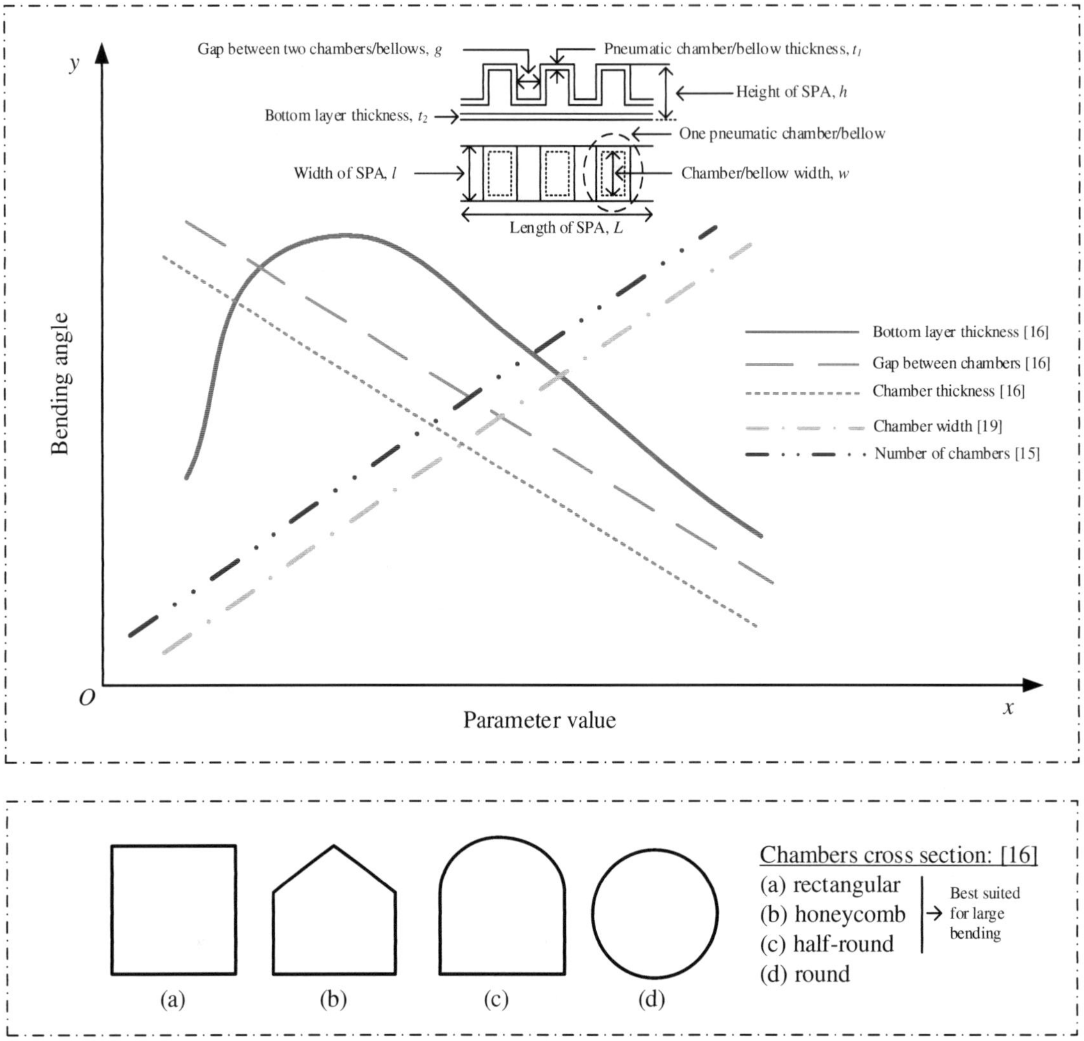

Figure 2. The influence of different geometrical parameters on SPAs bending performance

In a FEM simulation it is very important to have a proper material model. NinjaFlex® TPU cannot be modelled by using a constant modulus of elasticity because rubber-like materials deformations are nonlinear, quite large, and reversible. However, hyperelastic modelling can be used to overcome these problems. Reppel et al. [20] determined experimentally the elastic properties of printed NinjaFlex® TPU and fitted the experimental data on a 2nd order Ogden model. Ogden hyperelastic model for incompressible isotropic material under uniaxial stress can be defined as [21]:

$$W(\lambda) = \sum_{p=1}^{N} \frac{\mu_p}{\alpha_p}\left(\lambda^{\alpha_p} + 2\lambda^{-\frac{1}{2}\alpha_p} - 3\right) \tag{1}$$

where W is the strain energy density function, λ is the stretch, N is model order and μ_p and α_p are material coefficients.

If Eq. (1) is partially derived as a function of stretch λ, the uniaxial stress-stretch $\sigma(\lambda)$ relationship can be written as follows:

$$\sigma(\lambda) = \sum_{p=1}^{N} \mu_p \left(\lambda^{\alpha_p - 1} - \lambda^{-\frac{1}{2}\alpha_p - 1} \right) \tag{2}$$

In Chapter 4 the following coefficients will be used for FEM simulation of proposed SPAs made from NinjaFlex® TPU: $\mu_1 = 0.133$ MPa, $\alpha_1 = 3.05$, $\mu_2 = -1214$ MPa, $\alpha_2 = -0.0054$, D_1=D_2=0 MPa^{-1} (assumption of incompressibility [21]). The uniaxial stress-stretch characteristic $\sigma(\lambda)$ based on coefficients presented above is presented in figure 3.

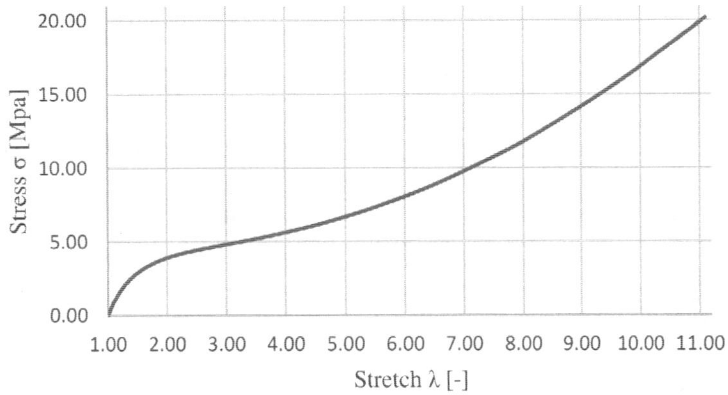

Figure 3. Uniaxial stress-stretch σ(λ) nonlinear curve of NinjaFlex® TPU for elongation up to 1100%

3. Bellow-type SPAs design

Based on the analysis presented in Chapter 2, two types of SPAs (*Actuator A1* and *Actuator A2*) are proposed in this paper. The numerical values of the main geometrical parameters are listed in table 1 while CAD models are presented in figure 4.

Regarding the cross-section type, both actuators have a half-round section but with different orientation. This choice was made to simplify the 3D-printing process for Actuator A1 especially which was designed to be printed with chambers/bellows in vertical position. In this way, the top layer can be printed using overhangs and the SPA can be obtained in one print. On the other hand, Actuator A2 was designed to be printed on its sideway (chambers positioned horizontally). This way of printing has some advantages as highlighted by Yap et al. in [10]. For example, the walls are printed longitudinally, thus in the direction of chambers deformation and not transversally like in the case of Actuator A1, but the overhangs area (the top sidewall) is much bigger compared with the one of Actuator A1 (the half-round top of the chambers). Actuator A2 can be manufactured from one print, too.

Geometrical parameters were chosen to account the nozzle size of the 3D printer (0.4 mm) because influences the layer height and wall thickness in the 3D-printing process. Both actuators have ten pneumatic chambers. The other geometrical parameters were slightly modified based on the bending performance considerations presented in Chapter 2 and to accommodate the dimension of the human pointing finger. The main difference between the two proposed SPAs designs is the way the chamber walls are deformed when they are pressurized. Actuator A1 permits the touching of two adjacent chambers walls when the pressure goes over a threshold, while Actuator A2 geometry does not permit the contact between walls. Therefore, the bending deformation should be greater in the case of Actuator A1 when the pressure hits a certain value. A hole of 3.5 mm was made on lateral face of each actuator to allow the insertion of a pneumatic tube/connector to pressurize the pneumatic chambers.

Based on these observations, the CAD models of both actuators were imported in Ultimaker CURA® slicer software and then printed on a Creality Ender 3 V2 3D printer using snow white NinjaFlex® TPU. The printing results are shown in figure 4. Although the 3D-printing process was successful for both SPAs, we encountered airtightness problems especially on layers printed with overhangs.

To analyze the bending performance and stress level of the proposed actuators for different input pressures, we implemented a FEM simulation in ANSYS® Workbench. The obtained simulation results will be discussed in the next chapter.

Table 1. SPAs geometrical parameters (please refer to Fig. 2)

Parameter	*L* (*mm*)	*l* (*mm*)	*h* (*mm*)	*g* (*mm*)	t_1 (*mm*)	t_2 (*mm*)	*w* (*mm*)	*n* (*-*)
Actuator A1	98.4	15	18.7	1.6	1.2	1.2	11	10
Actuator A2	79	15	13.3	1	1.6	1.6	10.2	10

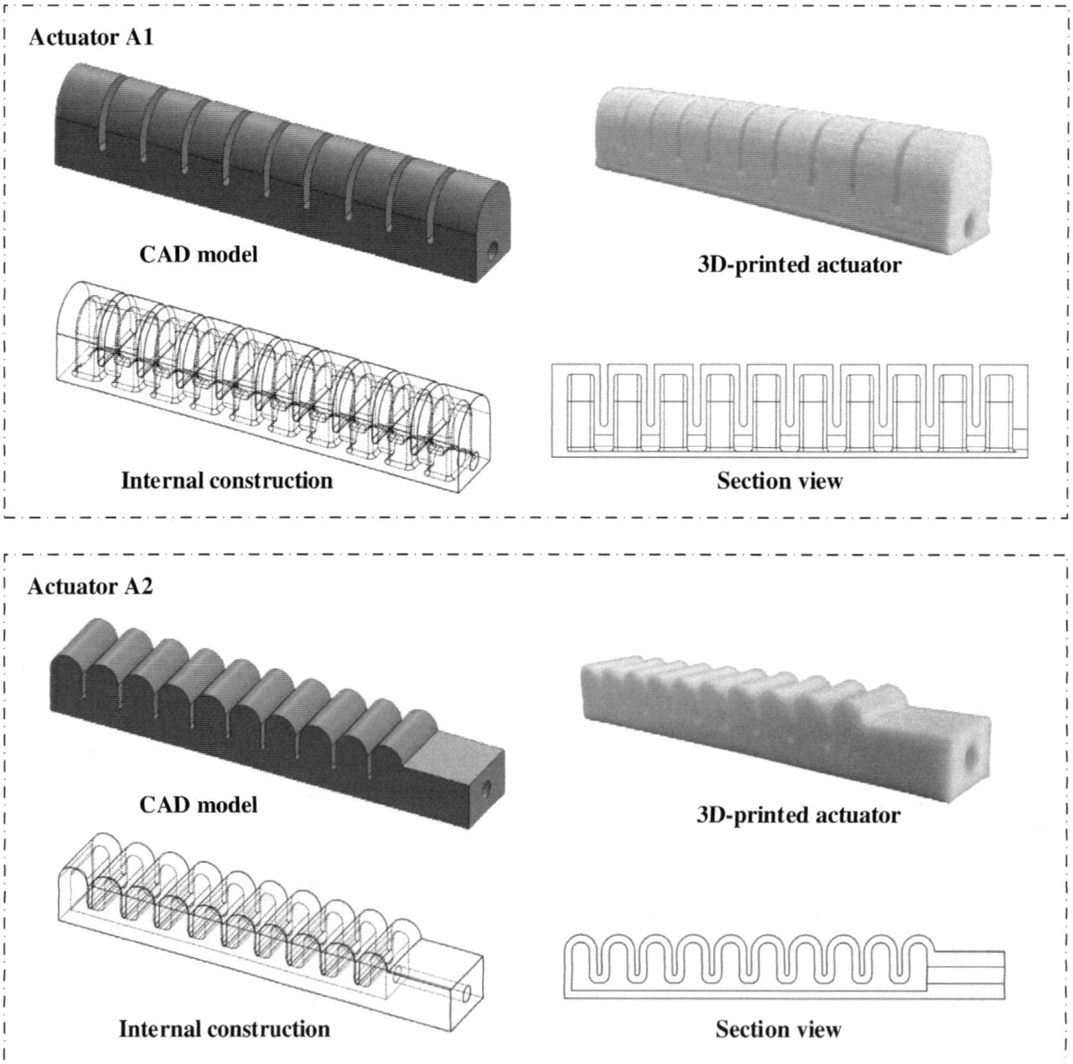

Figure 4. CAD models and 3D-printed versions of the proposed SPAs

IOP Conf. Series: Materials Science and Engineering 1268 (2022) 012012 doi:10.1088/1757-899X/1268/1/012012

4. FEM simulation and analysis

FEM simulation and analysis was implemented in ANSYS® Workbench. The following implementation steps were followed for both SPAs proposed in Chapter 3:

a. a new *Static Structural Analysis* project was created, and then the CAD model of the actuator was imported,
b. NinjaFlex® TPU material properties were defined using 2nd order Ogden model with the coefficients presented in Chapter 2 and material density was chosen to be 1190 kg/m^3 [14],
c. a coarse mesh was generated using higher-order tetrahedral elements with a size of 0.001 m (fine mesh is not recommended for large deformations as highlighted in [13]),
d. the boundary conditions (*Fixed Support* at one end of the actuator), loads (gravitational force and input pressure) and contacts (*frictional contact pairs* between surfaces that comes in contact when the deformation takes place) were defined,
e. *Large Deflection* option was selected to account for large deflection of the material,
f. FEM results were obtained and then evaluated.

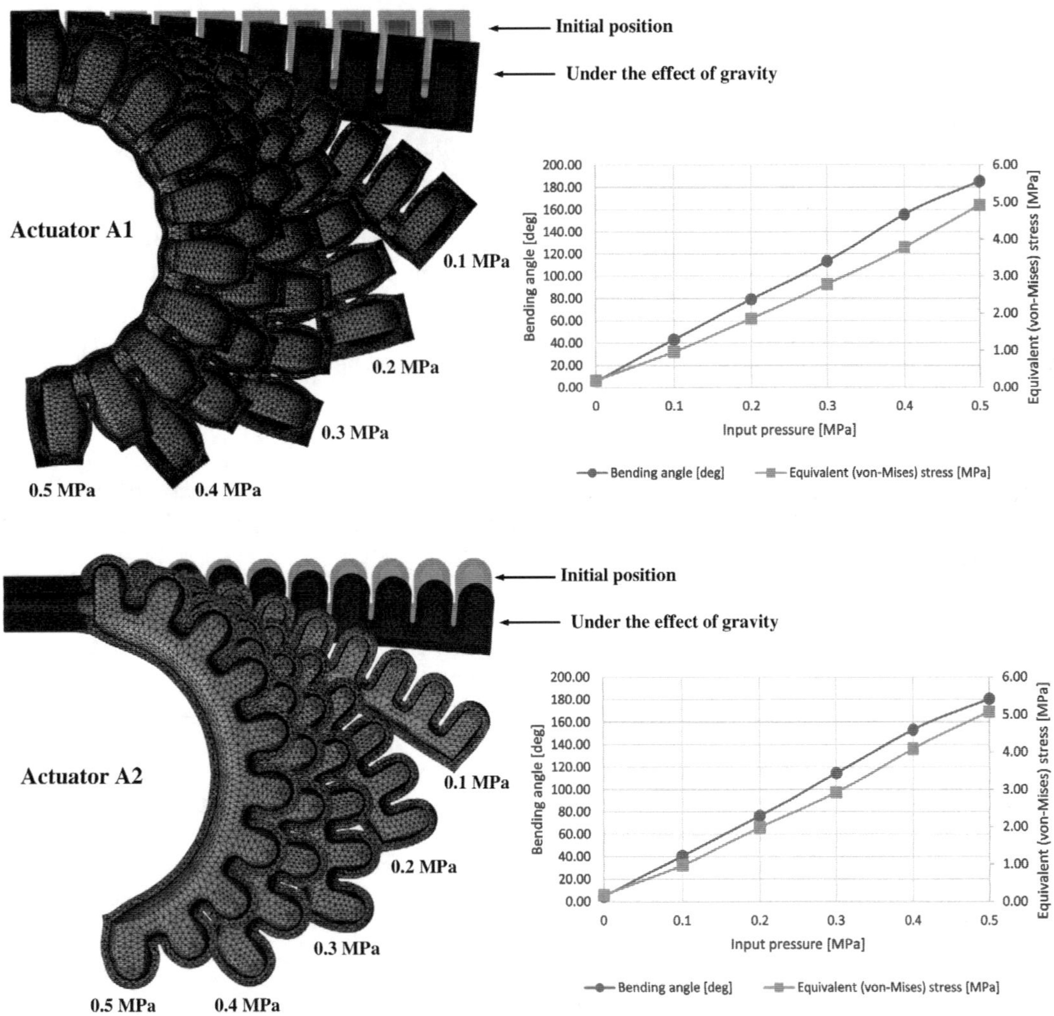

Figure 5. FEM simulation results for the proposed SPAs

The FEM results for the proposed SPAs are shown in figure 5. The actuators were first tested only under the effect of gravity to see how NinjaFlex® TPU material behaves when the chambers are not pressurized. Then, the bending behavior of the actuators was evaluated by also considering an input pressure between 0.1÷0.5 MPa with an increment of 0.1 MPa. For each input pressure, bending angle and equivalent (von-Mises) stress values were recorded and then used to obtain the characteristics presented in figure 5. Simulation result shows that both actuators generate relatively large deformations, but with a parametric design optimization procedure, the actuators could be designed to generate the same bending angle but at a lower input pressure.

5. Discussion

Based on the simulation results presented in Chapter 4, it can be seen that the bending performance of both actuators is very similar.

For the maximum input pressure of 0.5 MPa, Actuator A1 generates a maximum bending angle of 185.4 deg while Actuator A2 an angle of 180.54 deg. The difference of almost 5 deg is maintained also for other input pressures. Bending angle increases linearly with the input pressure for both actuators. This is because each chamber of the actuator has the same geometry and thus the curvature of the actuator is constant. That means that the shape of the actuator is a circular deformation curve as can be seen in figure 5. The bending angle vs input pressure characteristics can be modelled as a second order polynomial function as stated in [8,9]. The coefficient of the squared pressure term is inversely proportional with the modulus of elasticity of the material. NinjaFlex® TPU material has a modulus of 12 MPa which is much higher compared with other flexible materials used for fabrication of SPAs. Thus, in our case the relationship between bending angle and input pressure can be reduced to a linear one, based on the simulation results.

According to NinaFlex® TPU datasheet [14], the yield tensile strength of the material is 4 MPa and ultimate tensile strength is 26 MPa. If the characteristics between equivalent (von-Mises) stress and input pressure is analyzed (figure 5), both actuators can be pressurized at a maximum pressure of 0.4 MPa. After this point the elastic deformation of the material passes the yield point, the deformation of the material becomes irreversible, and the mechanical failure can occur.

6. Conclusions

The paper proposed a FEM simulation and analysis of two 3D-printed bellow-type bending SPAs made from NinjaFlex® TPU material. The design of the actuators was thoroughly investigated by considering the critical parameters (geometrical parameters and material type) that influences the bending deformation. 3D-printing manufacturing aspects were considered too.

As presented in figure 5, Actuator A1 permits the touching of two adjacent chambers walls when the pressure hits 0.2 MPa. Although the simulation results revealed a deviation of approximately 5 deg between the two proposed SPAs, the value represents a difference of only 3%, which in our case is not enough to say that the touching walls effect has a major impact on bending performance of Actuator A1 compared with Actuator A2. Regarding the 3D-printing process, the walls are printed longitudinally in the direction of deformation for Actuator A2 which is more advantageous. Also, the printing time was almost half in the case of Actuator A2 for the same printing speed parameters.

The obtained results can serve as a guideline to perform parametric design optimization of SPAs to achieve better performances in terms of bending capabilities. For example, the bending angle should increase proportional with the number of pneumatic chambers (considering that each chamber has the same geometry) when the other parameters remain unchanged. The shore hardness of the material is also important. NinjaFlex® TPU has a value of 85A but other ultra-soft materials such as FilaFlex® 60A TPU can be used in printing process. These aspects are planned to be investigated in future work.

7. References

[1] Lee C *et al.* 2017 Soft robot review *Int. J. Control Autom. Syst.* **15** 3–15

[2] Lapusan C, Hancu O and Rad C 2022 Shape Sensing of Hyper-Redundant Robots Using an AHRS IMU Sensor Network *Sensors* **22** 373
[3] Lapusan C, Rad C and Hancu O 2021 Kinematic analysis of a hyper-redundant robot with application in vertical farming *IOP Conf. Ser. Mater. Sci. Eng.* **1190** 012014
[4] Shintake J, Cacucciolo V, Floreano D and Shea H 2018 Soft robotic grippers *Advanced materials* **30(29)** 1707035
[5] Pagoli A, Chapelle F, Corrales-Ramon JA, Mezouar Y and Lapusta Y 2021 Review of soft fluidic actuators: classification and materials modeling analysis *Smart Materials and Structures* **31(1)** 013001
[6] Rad C, Hancu O and Lapusan C 2020 Aspects regarding "soft" grasping in smart agricultural harvesting tasks *Acta Tech. Napoc. Ser. Appl. Math. Mech. Eng.* **63** 389–394
[7] Xavier MS *et al.* 2022 Soft Pneumatic Actuators: A Review of Design, Fabrication, Modeling, Sensing, Control and Applications *IEEE Access* **10** 59442-59485
[8] Zolfagharian A, Parvez Mahmud MA, Gharaie S, Bodaghi M, Kouzani AZ and Kaynak A 2020 3D/4D-printed bending-type soft pneumatic actuators: Fabrication, modelling, and control *Virtual and Physical Prototyping* **15(4)** 373-402
[9] Rad C, Hancu O and Lapusan C 2022 Data-Driven Kinematic Model of PneuNets Bending Actuators for Soft Grasping Tasks *Actuators* **11** 58
[10] Yap HK, Ng HY and Yeow CH 2016 High-force soft printable pneumatics for soft robotic applications *Soft Robotics* **3(3)** 144-158
[11] Amiri Moghadam AA, Caprio A, Alaie S, Min JK, Dunham S and Mosadegh B 2019 Rapid Manufacturing of Thin Soft Pneumatic Actuators and Robots *J Vis Exp.* **8(153)** e60595
[12] Xavier MS, Fleming AJ and Yong YK 2021 Finite element modeling of soft fluidic actuators: Overview and recent developments *Advanced Intelligent Systems* **3(2)** 2000187
[13] Tawk C and Alici G 2020 Finite Element Modeling in the Design Process of 3D Printed Pneumatic Soft Actuators and Sensors *Robotics* **9** 52
[14] NinjaFlex® 3D Printing Filament: Technical Specifications. Available online: https://ninjatek.com/wp-content/uploads/NinjaFlex-TDS.pdf (accessed on 24 July 2022)
[15] Gao R, Fang J, Qiao J, Li C and Zhang L 2022 The bending control of the soft pneumatic finger *In Journal of Physics: Conference Series* **2181(1)** 012060
[16] Hu W, Mutlu R, Li W and Alici G 2018 A Structural Optimisation Method for a Soft Pneumatic Actuator *Robotics* **7** 24
[17] Wilson JF 1984 Mechanics of bellows: A critical survey *International Journal of Mechanical Sciences* **26(11-12)** 593-605
[18] Stano G, Arleo L and Percoco G Additive Manufacturing for Soft Robotics: Design and Fabrication of Airtight, Monolithic Bending PneuNets with Embedded Air Connectors *Micromachines* **11** 485
[19] Sun Y, Zhang Q, Chen X and Chen H 2019 An Optimum Design Method of Pneu-Net Actuators for Trajectory Matching Utilizing a Bending Model and GA *Mathematical Problems in Engineering* **2019** 6721897
[20] Reppel T and Weinberg K 2018 Experimental determination of elastic and rupture properties of printed Ninjaflex *Technische Mechanik-European Journal of Engineering Mechanics* **38(1)** 104-112
[21] Ćurković P and Jambrečić A 2020 Improving structural design of soft actuators using finite element method analysis *Interdisciplinary Description of Complex Systems: INDECS* **18(4)** 490-500

Acknowledgments
This work was supported by a grant of the Romanian Ministry of Education 250 and Research, CCCDI - UEFISCDI, project number PN-III-P2-2.1-PED-2019-4939, within PNCDI 251 III.
All authors contributed equally to this manuscript.

IOP Conf. Series: Materials Science and Engineering 1268 (2022) 012013 doi:10.1088/1757-899X/1268/1/012013

The influence of the cooling conditions and inclination of surfaces obtained by 3D printing on the roughness parameters

Adelina Hrițuc, Andrei Marius Mihalache, Laurențiu Slătineanu, Oana Dodun, Gheorghe Nagîț, Marius Ionuț Rîpanu

"Gheorghe Asachi" Technical University of Iaşi, Romania

Corresponding author's e-mail address: hrituc.adelina3295@yahoo.com

Abstract. A characteristic of the evolution of manufacturing processes in recent years is the expansion of technologies based on additive manufacturing, especially in the case of parts made of plastic materials. One of the accessible processes for making plastic parts is fused filament fabrication. In many situations, when using fused filament fabrication, it is necessary to obtain surfaces characterized by certain values of roughness parameters. Several factors can influence surface roughness. Some of these factors consider the materialization conditions of the 3D printing process. Starting from the hypothesis that the cooling conditions and the surface position can influence the roughness parameters values, experimental tests were carried out by modifying the values of the two factors. The experimental results were mathematically processed, and power function type mathematical models were determined for the roughness parameters *Rz*, *Rt*, *Rp*, *Ra*, *Rq*, *Rc*. These models provide information on the direction of action, the intensity of the influence exerted by the cooling conditions, and the angle of inclination of the test sample surface on the values of the mentioned roughness parameters.

1. Introduction

Manufacturing parts by 3D printing is included in a wider group of manufacturing processes called *additive manufacturing*. In principle, manufacturing by 3D printing involves a gradual realization of the part by depositing successive layers using processes that allow such deposition [1]. The development of 3D printing processes took place mainly in the last three decades when computer programs provided conditions for a certain slicing of the solid body of the part in its digital format.

Currently, additive manufacturing tends to spread more and more due to the simplification of the operating modes of 3D printing equipment and mainly due to the decrease in equipment price. One of the essential advantages of 3D printing results from the possibility of making parts of complex shapes without the need for the successive use of complex machining processes.

It should be emphasized that 3D printing is used to a greater extent to make parts from plastic materials, given their more appropriate properties to the requirements of the 3D printing manufacturing method.

There are several possibilities for classifying 3D printing methods. Suppose the principle underlying the successive addition of layers is taken into account. In that case, it is found that there are 7 groups of processes: vat photopolymerization, material extrusion, material jetting, binder jetting, powder bed fusion, direct energy deposition, and sheet lamination [2].

The manufacturing process with fused filament (FFF, fused filament fabrication) involves using a filament from the part material. The filament is advanced inside a nozzle where it melts, and the molten material is gradually placed on the printer table due to numerically controlled relative movements between the nozzle and the printer table.

The evaluation of 3D printing equipment can be carried out by taking into account its capabilities to achieve fine details, low roughness, parts resistant to different categories of demands, the productivity of the deposition process, etc.

The roughness of a surface is defined as an assembly of asperities of reduced height and for which the ratio between the length *s* and the height *h* of the asperities is less than 50 ($s/h<50$). The values of some roughness parameters evaluate the size of the roughness of a surface. Knowing the values of the roughness parameters is important for the maintenance over time of the character of the joints, wear, tightness, the value of the coefficient of friction, etc. There are 5 groups of such parameters.

If, in the case of cutting processes, the asperities are the result of the action of a cutting tool on the material of the workpiece, in the case of the fused filament fabrication method, the asperities result from the successive deposition of layers and the behavior of the material during deposition.

Investigating some aspects related to the roughness of surfaces made by 3D printing, in general, has been a concern of researchers in manufacturing.

Thus, the influence exerted by the temperature variation on the values of the roughness parameters *Ra*, *Rz*, *Rt*, and *Rsm* was investigated by Chaidas et al. [3]. One of the study conclusions was that an increase in the temperature used to melt the filament increases the value of the roughness parameters.

The results of research on how the values of the roughness parameter Ra are influenced by the direction along which the measurement is carried out were presented by Alsoufi and Elsayed [4]. As expected, a measurement along a direction perpendicular to the printing direction leads to more truthful information regarding the surface roughness of the test samples or parts manufactured by 3D printing.

Interesting in studying some mechanical and tribological properties of parts manufactured by 3D printing from polylactic acid incorporating bronze particles, Hanon et al. paid attention to the effect exerted by a print orientation by taking into account the roughness of the surfaces [5].

Experimental research aimed at the influence exerted by the direction of the relative movement between the nozzle and the part being generated and by the angle of inclination of the generated surface on the value of the roughness parameter Sa specific to some surfaces that can be achieved by nonplanar 3D printing was carried out by Elkaseer et al. [6]. They appreciated that a good quality surface could be obtained for values of the inclination angle lower than 25°.

Vidakis et al. aimed to investigate, among other things, the influence exerted by aluminum oxide particles (Al_2O_3) on the nanoscale surface roughness of some 3D printed samples of a nanocomposite containing polypropylene with different amounts of aluminum oxide nanoparticles [7]. The values of the roughness parameters *Rq*, *Ra*, and *Rz* were determined using a scanning electron microscope.

A finding formulated by Fico et al. [8] in a review article is that some properties of parts made of polymers are strongly influenced by the values of specific parameters, for example, of the fused filament fabrication process. Surface roughness, dimensional accuracy, and mechanical properties were included in the group of final properties.

In this article, the problem of the influence exerted by the angle of inclination of the surface and by the cooling conditions on the values of some roughness parameters related to the height of the asperities of some surfaces and respectively on the intensity of this influence was addressed.

2. Theoretical considerations

The FFF process involves the gradual advancement of a filament from the material to be deposited in a nozzle heated to controlled temperatures and moved along predetermined trajectories to the table of the 3D printer so that the material to be deposited is brought to a plastic state and thus disposed such that it forms layers made up of successively deposited filaments (figure 1, a). Upon exiting the nozzle, the filament to be deposited has a circular cross-section.

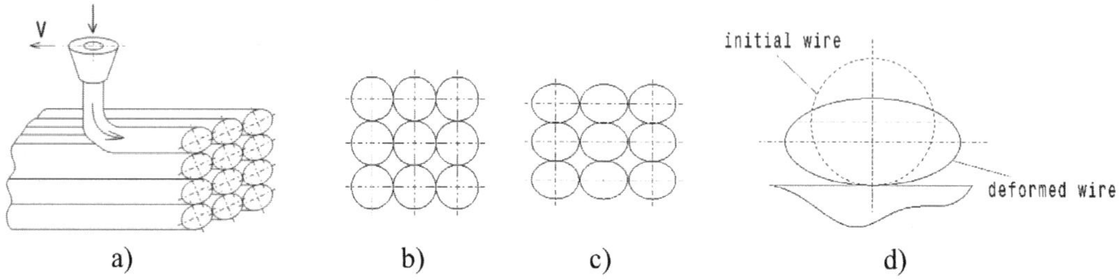

Figure 1. Deformation of the wire deposited by the FFF process: a – schematic representation of the wire deposition method; b – the shape of the deposited filaments if plastic deformation did not occur; c – the approximately elliptical shape of the filament in a cross-section; d – deformation of a filament from a circular cross-section to an approximately elliptical section.

If there were no processes of plastic deformation of the wire, layers of material deposited successively to obtain a test sample or part would highlight the circular shape of the filament in a section perpendicular to the direction of deposition (figure 1, b). In reality, due to the plastic state of the material of the wire leaving the nozzle, plastic deformation takes place, which means that in a section perpendicular to the deposition direction it can see the existence of an approximately elliptical shape of the wire embedded in the final sample (figure 1, c).

Analyzing how the wire is initially deposited on the table of the printing equipment, it can be accepted that due to its weight and the plastic state in which the material of the filament, the cross-section of the filament being deposited changes from an initial approximately circular shape to an approximately elliptical shape (figure 1, d).

At the part level, the profile of the surface generated by the successive deposition of plasticized material filament in a section perpendicular to the deposition direction could form a chain of ellipse arcs (figure 2).

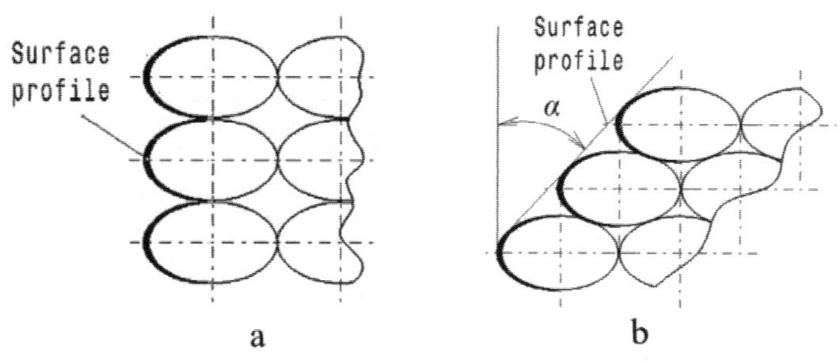

Figure 2. Generation of the profile of the surface of a part obtained by the FFF process: a – in the case of a vertical wall; b – in the case of a wall characterized by an angle of inclination α to the vertical direction.

Let's note that several groups of factors influence the ellipse profile that contributes to the generation of the asperities of the part surface:

- The physical-mechanical properties and the respective ones that characterize the thermal behavior of the wire material to be deposited;
- The temperature at which the wire is brought and extruded through the nozzle;
- The speed of relative movement between the nozzle and the 3D printer table;
- The degree and mode of cooling of the deposited wire, etc.

3. Experimental conditions

Through the experimental research, the aim was to obtain some empirical mathematical models that highlight the direction of action and the intensity of the influence exerted by the angle of inclination of the surface and by the cooling conditions in the FFF process on the values of roughness parameters of the surfaces of the polylactic acid (PLA) test samples manufactured by 3D printing. It was used to print a 3D part having the shape of a quarter of a disk, with the periphery made up of 6 zones with different inclinations to the vertical direction (figure 3). The shape of the part and the paths of relative movement between the nozzle and the printer table have been designed so that it becomes possible to highlight the influence exerted by the inclination of the surface to a vertical direction.

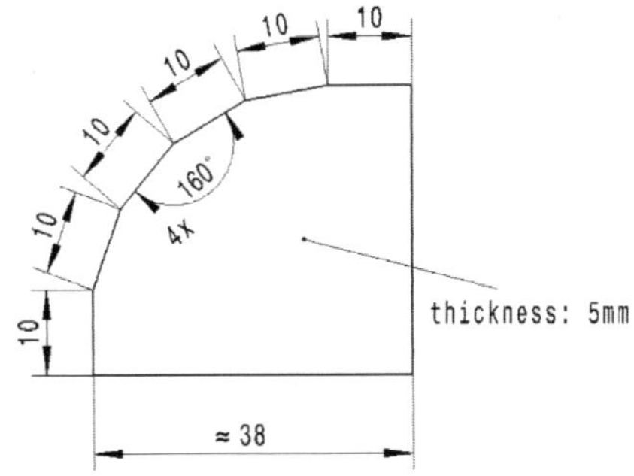

a

b

Figure 3. Sketch of the sample (a) and its image (b).

To make the samples, an Ultimaker 2+ type of 3D printing equipment (made by Ultimaker – Netherlands) was used. The values of the roughness parameters were determined using a roughness meter type Surtronic S25 (Taylor Hobson).

As an input factor in the printing process, apart from the angle α of inclination of the surface, the consideration of the cooling conditions c (in %) was established. The values of other input factors in the 3D printing process were kept constant (printing speed s=50 mm/s, layer height h=0.1 mm, infill density i=25 %, wall thickness t=0.7 mm, etc.). The values of some roughness parameters of the surfaces corresponding to the 6 zones arranged on the periphery of the quarter disc were measured with the help of a roughness meter type Taylor Hobson Surtronic S25.

Particular attention was paid to those roughness parameters that consider the heights of the existing asperities in the surface profile of the test samples. According to the standards in force, there are five groups of roughness parameters [2].

In this paper, values of some parameters from two of the five groups of roughness parameters were considered, namely:

a) From the group of amplitude parameters that follow aspects related to peak and valley:
- Maximum height of the assessed profile, *Rz*;
- Total height of the assessed profile, *Rt*;
- Maximum profile peak height of the assessed profile, *Rp*;

b) From the group of amplitude parameters that take into account the average of the ordinates:
- Arithmetical mean deviation of the assessed profile, *Ra*;
- Root mean square deviation of the assessed profile *Rq*,
- Mean height of the profile elements of the assessed profile *Rc*.

The selection of these roughness parameters for the experimental research was carried out considering the availability of the roughness meter and taking into account the hypothesis that the graphical representations of the variations could present similar aspects.

The values of the input factors (cooling conditions and the values of the angles of inclination of the surfaces) and the output parameters (the values of the roughness parameters *Rz*, *Rt*, *Rp*, *Ra*, *Rq*, and *Rc*) from the investigated process were entered in table 1.

The values of the angle of inclination (between 0 and 90 degrees) were established to obtain information regarding the influence of this factor on the values of the roughness parameters. When establishing the value of the factor that defined the cooling conditions, two values were taken into account of those commonly used in the 3D printing process, but sufficiently spaced apart.

Table 1. Values of the input factors and output parameters.

Meas. no.	Input factors		Output parameters, μm					
			First group			Second group		
	Cooling conditions, c %	Angle, α, ⁰	Rz	Rt	Rp	Ra	Rq	Rc
1	100	90	27,1	29,3	10,35	5,515	6,59	20,25
2	100	72	37,85	43,75	15,9	7,025	8,6	25,3
3	100	54	57,65	64,65	24,75	10,95	13,4	37,85
4	100	36	55,45	66,4	26,15	10,5	12,8	34,5
5	100	18	52,3	64,85	29,85	12,665	14,975	47,35
6	50	90	29,2	31,7	11,75	6,1	7,275	22,7
7	50	72	43,25	47,25	19,6	8,085	9,835	27,6
8	50	54	57,65	64,65	24,75	10,95	13,4	37,85
9	50	36	63,5	70,95	24,7	15,5	17,75	53,3
10	50	18	75,25	77,4	38,2	20,05	22,8	70,9

4. Processing of experimental results

For processing the experimental results, specialized software was used, developed by taking into account the least squares method [9]. The specialized program allows the selection of the most appropriate empirical mathematical model from five available models (first-degree polynomial, second-degree polynomial, power-type function, exponential function, hyperbolic function). The selection is made using the value of the so-called Gauss's criterion. This criterion involves taking into account the sum of the squares of the differences between the values determined with the help of the proposed model and, respectively, the values determined experimentally for the same values of the abscissas. The lower the value of Gauss's criterion, the more suitable the mathematical model is for the experimental results. The use of power-type functions was preferred because such functions provide direct information on the direction of action and the intensity of the influence exerted by an input factor on the value of the output parameter. The resulting equations using the specialized program and the values of Gauss's criterion for each of the equations are listed in Table 2.

Table 2. Empirical mathematical models and Gauss's criterion values S_G for each model.

Group	Equation	S_G
1	$Rp = 1073.19c^{-0.289}\alpha^{-0.688}$	28.50822
1	$Rz = 1241c^{-0.304}\alpha^{-0.501}$	111.9492
1	$Rt = 1026.33c^{-0.225}\alpha^{-0.55}$	126.7168
2	$Ra = 936.327c^{-0.448}\alpha^{-0.672}$	6.918129
2	$Rq = 863.345c^{-0.412}\alpha^{-0.645}$	8.687672
2	$Rc = 2965.006c^{-0.432}\alpha^{-0.662}$	77.73966

By considering the empirical mathematical models included in Table 2, the graphical representations in figures 4 and 5 were developed. The analysis of the mathematical models and graphical representations in figures 4 and 5 allowed the following observations to be formulated.

It is found that in the case of all 6 established empirical mathematical models, the increase in the values characterizing the cooling conditions and, respectively, the values of the inclination angle α of the surfaces obtained by 3D printing leads to a decrease in the values of the 6 roughness parameters considered. In all cases, a stronger influence is exerted by the angle α, since the values of the exponents attached to the factor α are higher than those of the exponents attached to the factor c.

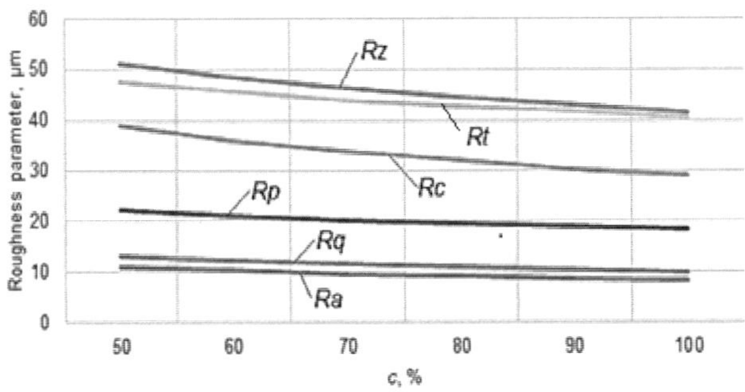

Figure 4. Influence exerted by the cooling conditions (%) on the values of some roughness parameters (α=54 °).

A possible explanation of the decrease in the values of the roughness parameters when increasing the value of the inclination angle could consider the change in the arrangement of the deposition filament with the elliptical section, as can be seen in figure 2, b. When increasing the value of the inclination angle, it is possible to greater filament penetration between previously deposited filaments. This results in a decrease in the values of the roughness parameters that take into account the heights of the generated surface asperities.

As for the decrease in the values of the roughness parameters upon the intensification of the cooling process, this could be explained by a faster solidification of the deposited filament, whose section failed to get closer to that of an ellipse. In principle, a profile made up of elliptical sections obtained by the deformation of some circular sections, will have higher values of the roughness parameters that take into account the heights of the asperities.

The variations are more pronounced in the case of the roughness parameters *Rz*, *Rt,* and *Rc* and are lower in the case of the roughness parameters *Rp*, *Rq,* and *Ra*.

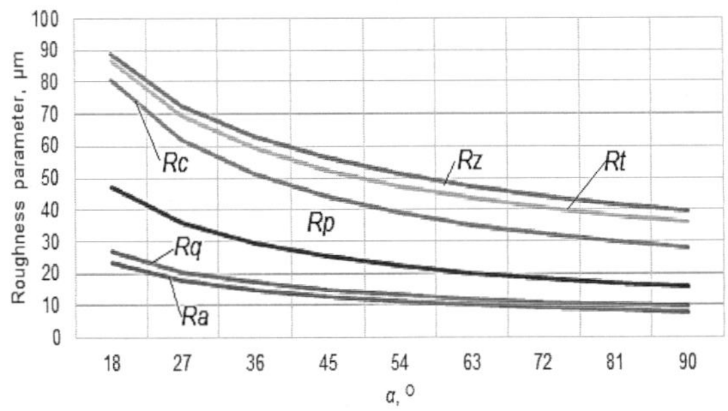

Figure 5. Influence exerted by the surface inclination angle α on the values of some roughness parameters (c=50%)

5. Conclusions

The influence of different factors on the values of surface roughness parameters in the case of 3D printed parts is an important issue. In the case of the fused filament fabrication process, it is expected that due to the plastic deformation of the molten material of the filament, there will be different values of the roughness parameters on the surfaces of the produced test sample. To obtain additional information regarding these aspects, experimental research was carried out based on the variation of the cooling conditions and, respectively, on the variation of the angle of inclination of the surface. Mathematical models of the power function type were determined through the mathematical processing of the experimental results. These models revealed the decrease in the values of the roughness parameters considered when the intensity of the cooling process increases and, respectively, when the angle between the realized surface and a vertical direction increases. It was found that a stronger influence is exerted by the value of the angle of inclination of the surface of the specimen. In the future, it is intended to expand the experimental research by considering other polymeric materials and using test samples of more complex shapes.

References

[1] Slătineanu L, Nagîț G, Dodun O, Coteață M, Munteanu A, Beşliu-Băncescu I, Gherman AL, Hrițuc A, Chinesta F, Goncalves-Coelho A, Teixeira J P, San Juan M, Santo L, Santos F 2020 *Electrophysical and chemical manufacturing processes* (Chişinău: Publishing House Tehnica Info)

[2] Manufactur3D. 2018 *The 7 types of additive manufacturing technologies* Available: https://manufactur3dmag.com/7-types-additive-manufacturing-technologies/ (Accessed: 5 07 2022)

[3] Chaidas D, Kitsakis K, Kechagias J, Maropoulos S 2016 *IOP Conf. Ser. Mater. Sci. Eng* 161 012033.

[4] Alsoufi MS, Elsayed AE 2017 *Am. J. Mech. Eng.* 5 211-222.

[5] Hanon MM, Alshammas Y, Zsidai L *Int. J. Adv. Manuf. Technol.* 108 553–570

[6] Elkaseer A, Müller T, Rabsch D, Scholz SG 2020 *Proceedings of the Sustainable Design and Manufacturing (Singapore)* p 285

[7] Vidakis N, Petousis M, Velidakis E, Mountakis N, Fischer-griffiths PE, Grammatikos SA, Tzounis L 2022 *Polym. Test.* 109 107545

[8] Fico D, Rizzo D, Casciaro R Corcione C E 2022 *Polymers* 14 465.

[9] Cretu G 1992 *Fundamentals of Experimental Research. Laboratory Handbook* (In Romanian) (Iasşi, România: "Gheorghe Asachi" Technical University of Iaşi).

IOP Conf. Series: Materials Science and Engineering 1268 (2022) 012014 doi:10.1088/1757-899X/1268/1/012014

Reconstruction of Orbital Wall by Using Hybrid Manufacturing Operations

Sergiu Pascu[1], Nicolae Balc[1], Adrian Augustin Pop[2]

[1]Dept. of Manufacturing Engineering, Technical University of Cluj Napoca
[2]Dep. of Electrical Machines and Drivers, Technical University of Cluj Napoca

E-mail: Sergiu_Pascu2002@yahoo.com

Abstract. This paper will discuss the solutions and the main parameters used to print a mold that will be used in a manufacturing process for a personalized implant in orbital fracture situations. The main purpose of this article is to rebuild the damaged orbit. The new orbit should have dimensions and accuracy similar to one of the original elements. Based on the computed tomography (CT) of the patient, the images are extracted and processed. With that information, a customized implant is manufactured. In this article, a case study of a patient suffering from an orbital fracture is presented. Based on the CT images, a mold was printed on additive manufacturing equipment (AM). The manufacturing parameters being highly correlated with the requirements that are needed for this case such as hardness, accuracy, surface roughness and support removal were chosen based on a series of tests and results from other studies. Another aspect that was analysed was related to the material from which was manufactured the mold. The forming and cutting of the mesh are simplified, by introducing the formed metallic mesh making it ready to be applied to real patients and therefore decreasing the effective time required for this kind of operation.

1. Introduction

In recent years, there has been considerable interest in using 3D printing in important areas such as medical research, the aerospace industry and machine building. 3D printing, also called rapid prototyping (RP) or additive manufacturing, (AM) represents the construction of a three-dimensional object from a CAD model or a digital 3D model [1]. Three-dimensional printing - 3D technology has become widely used in the clinical area, enabling more accurate reconstructions of human parts [2].

Medical applications where 3D printing is widely used are fields such as orthopedics and plastic surgery, which include maxillofacial surgery. This paper will present one technique that can be used in blow-out fractures to perform rehabilitation of a damaged orbit, which is a very complex surgery to recover its original shape and functionality [3][4]. Orbital bone damage influences the position and functionality of the intraorbital element, prompting issues like diplopia and visual sharpness aggravations [5]. To avoid these possible problems, it is important to precisely establish the possible inaccurate areas and fix the pre-injury [6]. The main objective is to give an understanding of how the new orbit was replicated for precise medical procedures. The second objective of this research was to present optimized parameters for the printing and milling of the orbit using hybrid manufacturing technology. Repairing the orbital fracture through the classical methods is considered a delicate issue due to the disadvantages of the manufacturing materials and the characteristic process errors that might occur during the manufacturing operation.

Another concern is the accuracy during the realignment of the old and new elements which are very sensitive. Joining and adjusting small, delicate and thin bone sections is typically impossible. For this reason, the type of material used in medical implants can vary from well-known materials to experimental ones. Materials like silicone, titanium, polyurethane foam, polylactic acid and 3D printed biomaterials are accessible to recreate human body elements as orbital walls whenever the situation requires [7][8].

In this paper, we aim to improve the process manufacturing of the implants, for a randomly chosen case study, by facilitating and decreasing the total time spent in surgery, reducing the error that can appear during the surgery when the mesh is formed and increasing the overall quality of the process through hybrid manufacturing operations.

2. Case Report

In 2019, an 8-year-old girl had an accident after falling from a tree. She was brought to the hospital with suspicions of orbital contusions. After performing a CT scan, the diagnosis showed a fracture of the left orbit which is illustrated in *Figure 1.*

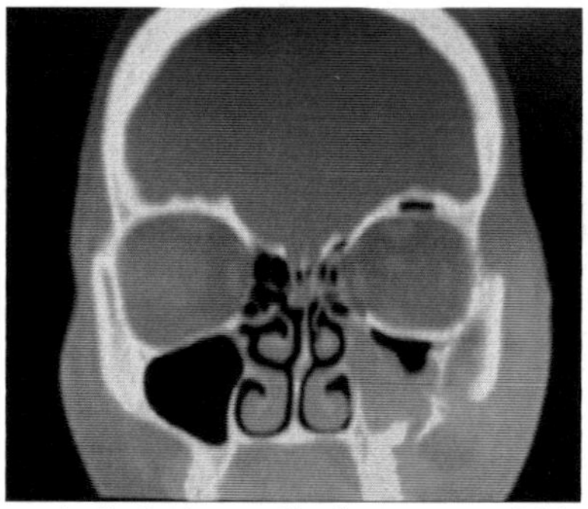

Figure 1. Soft tissue displacement within the maxillary sinus on CT scan

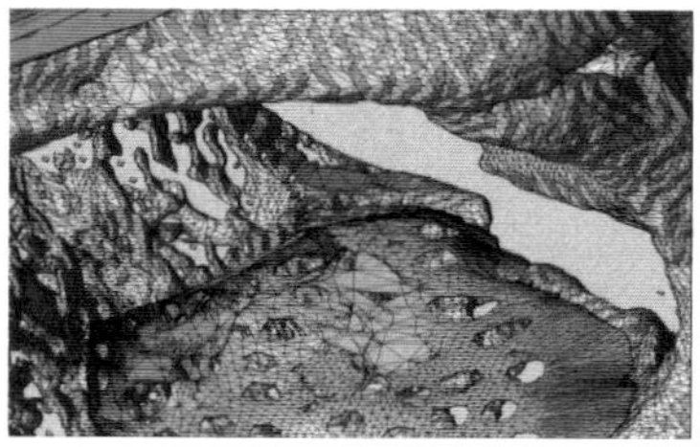

Figure 2. Orbit Anatomy- Alignment elements

The CT axial source images were post-processed in Vitrea software which is an advanced visualization system. This software is used mainly to improve the quality of the area that is targeted. This can be achieved by delivering a 3D model with high resolution. Similar 3D models of other patients are used for comparison purposes. Guidelines such as CT scans, printing strategies and the parameters describing orbital reconstruction using 3D modeling and segmentation using CT base images were found and used as a starting point for this particular case study [9]. The area showing the broken bone was isolated, the focus was showing only the orbits in the horizontal plane with the rear boundary identified as the rear edge of the optical channel. The quality control objective of this stage was to analyse the three planes: horizontal, coronal and sagittal. It is important to decide if the model that was created is an accurate representation of the images received from the CT source. The stereolithography (STL) model was divided into the same three symmetrical planes. This was done at the modeling stage to confirm the accuracy of the model while viewing the model in other 3D design software. Using the solution proposed, the next step to be performed was the cleaning of the model. The higher quality of the .STL file will result in a higher number of triangles as shown in *Figure 2.* The time for loading the surface will increase and more computationally resources will be needed.

Finishing with these steps of processing the images, the .STL file was loaded in the Blender 3D computer graphics software. This toolset is used for creating 3D printable models. To verify and compare the accuracy of the model within these two software, as was specified in the previous step, the model was measured in the exact same location as prior in Vitrea.

Every step made is realized in such a way that it will bring us closer to the desired goal. After performing a second verification of the model and the desired accuracy is achieved, the model is validated. In *Figure 3* two disjointed orbital CAD models are presented.

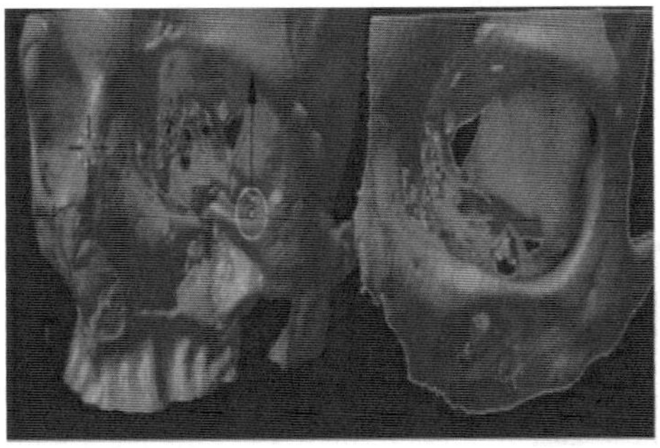

Figure 3. Both orbits before the overlapping

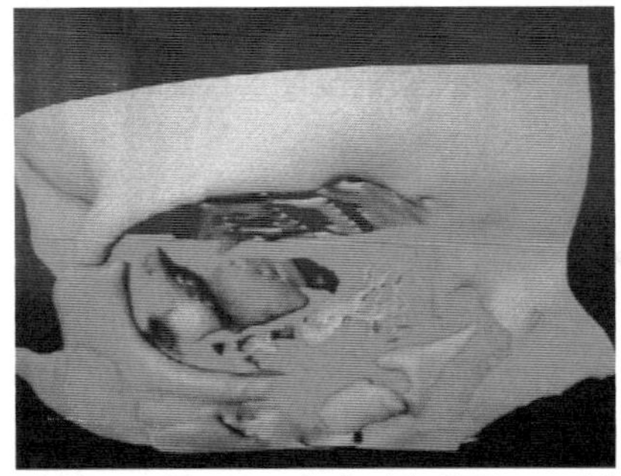

Figure 4. Fracture left side STL model

The starting point of designing the mold was to use a mirror tool in the Blender application. Through this tool, the mirroring of the healthy orbit right side was overlayed on top of the damaged orbit on the left side. The result is illustrated in *Figure 4*. The main reason for using the mirror tool on the midsagittal plane, even though the symmetry between orbits is not 100% as can be seen in *Figure 4*, was to obtain a model for the forming process of the mesh that is as similar as it can be to the damaged orbit before fracture. Even if a great symmetry between the models was obtained, the alignment was possible by taking into account the position of the elements that are illustrated in *Figure 5* and which are: (1) the optical channel, (2) the lateral orbit wall, (3) medial orbit wall, interior orbit wall and (4) anterior margin of the orbit.

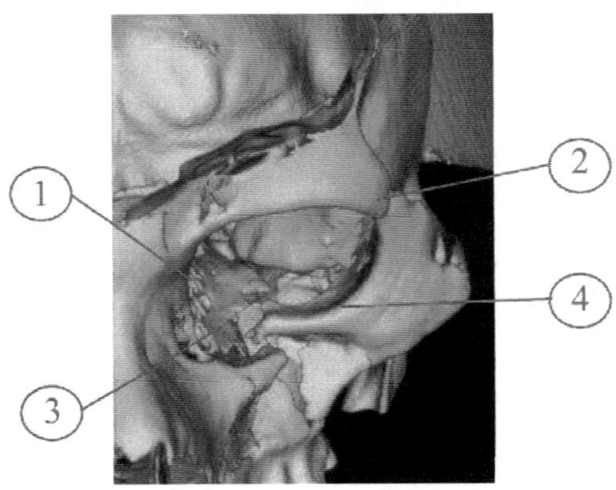

Figure 5. Orbit Anatomy- Alignment strategy

Once the overlapped model was obtained, a visual check was made followed by another series of measurements in the Blender software as in the steps presented above. This action was performed to see how high the variation between the damaged orbit is and the new overlapped one. For our case study, the variation was very small, less than 0.1 mm, taken as an average and measured in different critical areas. The model was saved in a .STL format and then imported into the Cura slicing application where the printing parameters will be established [10]. The most important aspect related to the process of printing parts and using them in the medical field is to obtain a very accurate physical model as compared to the original model. Once the model is printed, the metallic mesh is introduced into the orbit and is subjected to the forming process. It is very hard to modify the newly formed part after the process ends and especially during surgery. Taking these aspects into consideration but also analysing what other researchers did [11], it was concluded that a series of requirements is very important. These requirements are hardness, accuracy, surface roughness and support removal. To achieve the condition of hardness, the starting point was the results of another group of researchers [12] where 2 different materials PLA (Polylactic Acid) and ABS (Acrylonitrile Butadiene Styrene) were tested under the same conditions. The decision of going further with PLA as a printing material is because this material presents a 4.107GPa Young modulus, a tensile strength of 62.701MPa, elongation at break of 1.841% and toughness of 93.9MPa%. The sample that was tested to obtain these results had an infill of 100%. This infill will be also used in our research for the printing process. [13]

One of the best printing temperatures for PLA wire was found to be close to 215°C [14]. The filament used was FilaFill premium PLA, with a wire thickness of 1.75 mm and a melting point between 195°C-215°C. The same group of researchers used the following printing parameters: 90 mm/s infill speed, 100% infill density, 215°C nozzle temperature, 0.3556 mm layer thickness and a bed temperature between 23°C and 70°C. Taking these parameters as a baseline, the next step was to print a series of samples. The printing parameters were adjusted according to the printer performance used in this research which was a Prusa I3 with a single head. The parameters that provided good results were printing speed of 70 mm/s, nozzle size of 0.4 mm, layer height of 0.2 mm and printing temperature of 215°C. The nozzle size was chosen as 0.4 mm because it was the smallest nozzle available for this research. A bigger nozzle for instance 0.6 mm increases the surface roughness of the part.

The accuracy will be improved not just by using the parameters below, but also by correlating the flow with the temperature of the nozzle and the building supports. Another important aspect is related to the roughness of the part which should be lower than RA 20 µm. Values above this limit might produce misalignment between bone and metallic mesh, where RA is the roughness average. After a series of tests and also taking into consideration the approach that other researchers did[15]-[16], the parameters used for the building of the supports are presented in *Table 1*:

Table 1. Building Supports Parameters

Parameter	Value
Enable Conical Support	On
Conical Support Angle	45°
Conical Support Minimum Width	9 mm
Support Distance	1.2 mm
Support Interface Thickness	0.8 mm
Support Interface Density	80%
Support Roof Density	80%
Support Roof Line Distance	0.55 mm
Support Floor Density	80%
Support Interface Pattern	Concentric
Support Placement	Everywhere
Support Overhang Angle	55°
Support Pattern	Zig Zag
Support Density	12%
Support Z Distance	0.3 mm

Another aspect taken into consideration, was obtaining strong structural supports that can sustain the orbit in the printing process and which can be removed very easily leaving a good surface roughness. As the printing process was finished and the removal of the support's operation was completed, the roughness was measured in the area where the supports were removed. The values are presented in *Table 2* and the roughness curve is illustrated in *Figure 6*. The equipment used to measure the roughness within the specified area is a Mitutoyo Contracer CP - 200 / 400 - Contour Measuring Instrument.

IOP Conf. Series: Materials Science and Engineering 1268 (2022) 012014 doi:10.1088/1757-899X/1268/1/012014

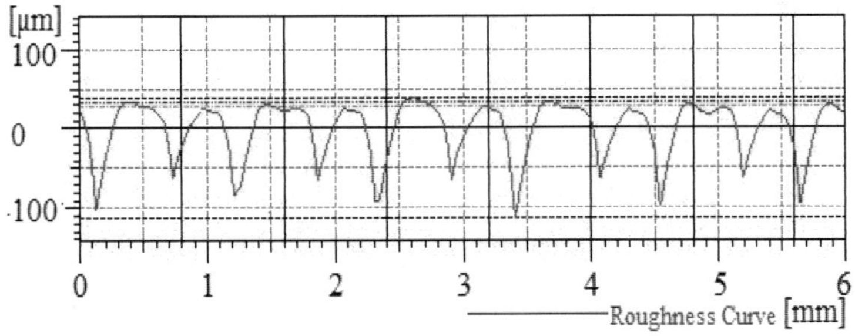

Figure 6. Roughness curve measurement before milling

As the value of the surface roughness exceeds the thresholds of Ra> 20 µm, it was necessary to apply a finishing operation to those areas.

Table 2. Roughness results before milling

Ra	26,9335 µm
Ra max	30,9907 µm
Ra min	20,1456 µm
Rasd	30,1576 µm
Ra+sd	30,0910 µm
Ra-sd	29,7759 µm
Ra (1)	30,9379 µm
Ra (2)	26,2322 µm
Ra (3)	29,1487 µm
Ra (4)	24,7410 µm
Ra (5)	30,9907 µm
Ra (6)	26,9545 µm

For the finishing process of the part, a Saeyang Marathon N3 high precision electric mill was used. The milling tool used was a Ø3 mm end mill. The nominal speed was 1.200 RPM, a value chosen based on the specification of the producer correlated with the values obtained from another group of researchers who used a Ø6 mm end mill with 3.500 RPM [17]. To assure the proper removal of the material and to check the quality of the surface, the roughness was measured again in the same area as previously. The results are presented in *Table 2*.

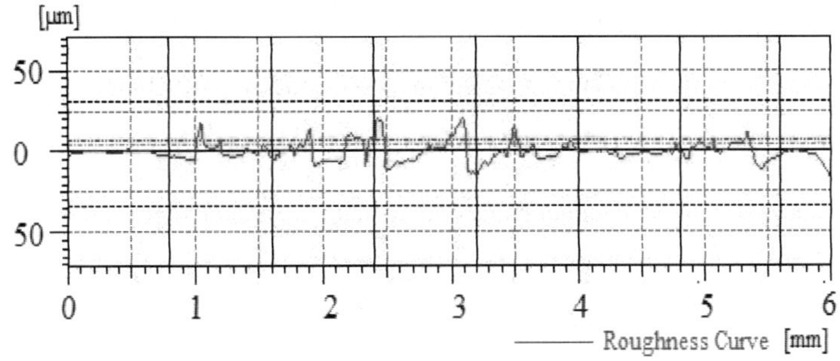

Figure 7. Roughness measurement curve after milling

As it can be seen in *Table 3* after the finishing operation, the roughness improved significantly from Ra 26,9335 µm to RA 5,111 µm.

Table 3.Roughness results after milling

Ra	5,111 μm
Ra max	11,135 μm
Ra min	0,855 μm
Rasd	2,898 μm
Ra+sd	8,0177μm
Ra-sd	2,2198μm
Ra (1)	0,855 μm
Ra (2)	3,355 μm
Ra (3)	5,696 μm
Ra (4)	1,461 μm
Ra (5)	4,188 μm
Ra (6)	1,461 μm

The newly obtained roughness assures a proper forming operation of the titanium mesh in the orbit. The last step of this operation was the forming process of the titanium mesh. Cobalt and titanium alloys are very used in the medical area especially in restoring orbital structure [18]. Cobalt alloys appear to be not so promising due to their poor functioning in orbital surgery and therefore they were gradually replaced by titanium implants[19] that became accredited by the Food and Drug Administration in 1984. Now the titanium mesh could be widely widespread in lots of applications as craniomaxillofacial surgery, particularly in massive defects. Titanium is chemically similar to calcium and so it is tissue tolerant. It is corrosion-resistant due to the spontaneously forming oxide layers at the surface. This ensures that the material behaves passively and there is no possible allergic reaction that can occur.

Computer-assisted designed and manufactured titanium implants have resulted in the most excellent reconstructive surgery of very complex elements like an optical nerve in safe conditions [20]. Small drawbacks of the Titanium mesh implant from a functionality point of view are caused by the abnormal edges of the mesh. This could apply pressure on smooth and fibrous tissue, which could make the implant difficult to change [21]. These generally utilized implants are created in big and small sizes which will be chosen based on the lateral orbital dimension, sex of the patient and age. From a review of anatomical studies, it was chosen to use the smaller implant and trim the area that was in excess [22].

In *Figure 8* it is illustrated that the implant might be inserted easily in the printed model due to the sleek edges.

Figure 8. Titanium mesh placed in the orbit

Depending on the size of the orbital wall, the tendency is to use bigger implants with no trimming operation performed additionally. After the mesh is inserted, it will be pressed against the edge of the orbit until the desired shape is achieved. Once the titanium mesh is formed correctly, it is mandatory to apply the proper sterilization method.

In this particular case study, gamma irradiation and H_2O_2 gas plasma treatment were used, which did not alter significantly the surface characteristics and chemical composition of the Zn or Zn alloy [23].

Once all the steps described above are done, the titanium mesh is introduced in the real orbit of the patient, details of fixing it will be established during the surgery.

3. Results Interpretation

The printed orbit was analyzed from the point of view of accuracy and quality of the surface. Optimized parameters resulted from intensive testing of the manufacturing operation considering all the variables, but also from other researchers. The 3D model of the orbit presents good quality of the surface, no signs of delamination of the layers and good behavior of the mold related to the strength of the part, when the mold was involved to the forming operation of the titanium mesh. As shown in *Figure 9*, the interior cavity and the complete reconstruction of the damaged orbit after printing and finishing operation presented good quality. The results can be seen in *Table 3*.

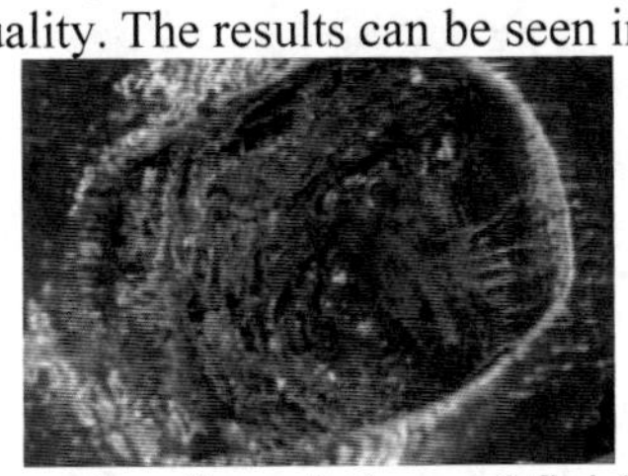

Figure 9. Interior cavity after printing and finishing operation

The roughness which was achieved through printing was 26.9 μm, higher than the value that was targeted, which was RA 20 μm. For better roughness, a finishing operation was applied, obtaining a new surface roughness of RA 5.1 μm. The measurements were performed inside of the orbit where the supports were in contact with the part.

The progress beyond the state-of-the-art can be observed by comparing the classic type of surgery in which the titanium mesh was formed manually and took around 100 minutes. The one in which a 3D mold was used, a decrease in surgery time of around 60 mins was noticed. The time used to create the model in Vitrea was around 32 mins. The modelling of the .STL file took around 65 mins. Each of these time estimations included measurements of the model on every step, as was presented in the case study.

The model was printed in 6 hours. Washing the model followed by a drying operation took 35 mins. Removing the excess material from both elements mold and titanium mesh took approximately 45 mins. Sterilization of the elements took around 35 mins. The total time consumed to achieve these steps was around 12 hours.

4. Conclusions

Computer-assisted surgical tools were used in this case study to significantly diminish the time required for this type of surgery. The most feasible solution for this case study was to use a 3D printed model of the orbit in combination with milling operations in order to obtain a higher surface finish. This approach may be utilized to create molds used in the forming process, a solution that can be used in forming the bioabsorbable implants utilized in custom-designed surgical repair. The resulted benefits were decreasing tissue handling, decreasing the surgery time and avoiding postoperative edema. This type of approach can be used and adapted for other types of medical conditions in the future. Future researchers should certainly further test whether the manufacturing time of the orbit can be optimized by developing procedures customized for each phase.

5. Acknowledgment
This research was financially supported by the Project "Entrepreneurial competencies and excellence research in doctoral and postdoctoral programs -Antredoc", a project co-founded by the European Social Found financing agreement no.56437/24.07.2019

References

[1] ASTM F2792•12a, Standard Terminology Additive Manufacturing Technologies ASTM International, West Conshohocke PA, 2012

[2] Daniel J. Thomas, Iain S. Whitaker, Zita M. Jessop, *3D Bioprinting for Reconstructive Surgery*, Techniques and Applications 2017

[3] D. Stanco, P. Urban, S. Tirendi, G. Garlli, J. Barrero, 3D bioprinting for orthopedic applications: current advances, challenges, and regulatory considerations, Bioprinting 2017

[4] R. Krishna, A. Manmadha, P. Bal Reddy, S. Vekatesh, Development of optimum preplanning for maxillofacial surgery using selective laser sintering, Int. J. Eng. sci. Technol. 3 (2011) IS-**196**.

[5] S. Manolidis, B. H Wæks, M. Kirby, M. Scarlett, L Hollier, Classification and surgical management of orbital fractures: experience with I I I orbital reconstructions, J. Craniofac. Surg. **13** (2002) 726-738.

[6] B. Hammer, J. Prei Correction of post traumatic orbital deformities: operative techniques and patients, maxillofacial Surg. **23** (1995)

[7] Myeroff Chad, Archdeacon Michael, Autogenous Bone Graft: Donor Sites and Techniques, *The Journal of Bone & Joint* December 7, 2011 - **93** - Issue 23 - p 2227-2236

[8] R.Schön, M.C.Metzger, C.Zizelmann, N.Weyer, R.Schmelzeisen, Individually preformed titanium mesh implants for a true-to-original repair of orbital fractures, *International Journal of Oral and Maxillofacial Surgery* **35**, Issue 11, November 2006, Pages 990-995

[9] Vehmeijer M, van Eijnatten M, Liberton N, Wolff J. A novel method of orbital floor reconstruction using virtual planning, 3-dimensional printing, and autologous bone, *Oral Maxillofac Surg* 2016;**74**:1608–12.

[10] Bell RB, Markiewicz MR. Computer-assisted planning, stereolithographic model-ing, and intraoperative navigation for complex orbital reconstruction: a descriptive study in a preliminary cohort. *Oral Maxillofac Surg* 2009;**67**:2559–70.

[11] Weijian Huaa, Weiliang Shib ,Kellen Mitchella, Lily Raymonda, Ryan Coultera, Danyang Zhaob, Yifei Jina, 3D Printing of Biodegradable Polymer Vascular Stents: A Review, *Chinese Journal of Mechanical Engineering*: Additive Manufacturing Frontiers Volume **1**, Issue 2, June 2022, 100020

[12] Adrián Rodríguez-Panes, Juan Claver and Ana María Camacho, The Influence of Manufacturing Parameters on the Mechanical Behaviour of PLA and ABS Pieces Manufactured by FDM: A Comparative Analysis, Department of Manufacturing Engineering, Universidad Nacional de Educación a Distancia (UNED), Madrid 28040, Spain

[13] Luca Fontana Paolo Minetola Luca Iuliano Serena Rifuggiato Mankirat Singh Khandpur Vito Stiuso, An investigation of the influence of 3d printing parameters on the tensile strength of PLA material, Volume **57**, Part 2, 2022, Pages 657-663

[14] Chamil Abeykoon, Pimpisut, Sri Amphorn, Anura Fernando, Optimization of fused deposition modeling parameters for improved PLA and ABS 3D printed structures, International Journal of Lightweight Materials and ManufactureVolume **3**, Issue 3, September 2020, Pages 284-297

[15] Luca Fontana Paolo Minetola Luca Iuliano Serena Rifuggiato Mankirat Singh Khandpur Vito Stiuso, An investigation of the influence of 3d printing parameters on the tensile strength of PLA material, Volume **57**, Part 2, 2022, Pages 657-663

[16] Lokesh N.Praveena B.A.Sudheer Reddy J. Evaluation on the effect of printing process

IOP Conf. Series: Materials Science and Engineering 1268 (2022) 012014 doi:10.1088/1757-899X/1268/1/012014

parameter through Taguchi approach on mechanical properties of 3D printed PLA specimens using FDM at constant printing temperature, Volume **52**, Part 3, 2022, Pages 1288-1293

[17] Răzvan Gabriel Pămărac, Petrusudy Regarding the Optimal Milling Parameters for Finishing 3D Printed Parts from ABS and PLA Materials, **70** (2018),66 – 72.

[18] Ling-Xiao Ye , Xiao-Ming Sun , Yu-Guang Zhang , Ying Zhang, Materials to facilitate orbital reconstruction and soft tissue filling in posttraumatic orbital deformities, Department of Plastic and Reconstructive Surgery, Ninth People's Hospital affiliated to the Medical School of Shanghai Jiao Tong University, Shanghai 200011, China.

[19] Schubert W, Gear AJ, Lee C, Hilger PA, Haus E, Migliori MR, Mann DA, Benjamin CI. Incorporation of titanium mesh in orbital and midface reconstruction. Plast Reconstr Surg 2002;**110**:1022-30; discussion 1031-2.

[20] Lieger O, Richards R, Liu M, Lloyd T. Computer-assisted design and manufacture of implants in the late reconstruction of extensive orbital fractures. Arch Facial Plast Surg 2010;**12**:186-91.

[21] William J. Weadock, Curtis J. Heisel, Alon Kahana, Use of 3D Printed Models to Create Molds for Shaping Implants for Surgical Repair of Orbital Fractures 2020 Apr;27(4):536-542.

[22] Chepelev L, Wake N, Ryan J, et al. Radiological Society of North America(RSNA) 3D printing Special Interest Group (SIG): guidelines for medical3D printing and appropriateness for clinical scenarios. 3D Print Med2018.

[23] Ping Li, Wentai Zhang, Sebastian Spintzyk, Impact of sterilization treatments on biodegradability and cytocompatibility of zinc-based implant materials, *Materials Science and Engineering*: Volume **130**, November 2021, 112430

Criteria for analyzing online traffic encountered in web and mobile applications in order to model an automated application testing system

C Duta[1], N Carutasu[2], I Stanica[3] and A Georgian[4]

[1]Doctoral School of Industrial Engineering and Robotics, Faculty of Industrial Engineering and Robotics, University Politehnica of Bucharest, Splaiul Independenţei 313, Bucharest, Romania.
[2]Department of Engineering and Management, Faculty of Industrial Engineering and Robotics, University Politehnica of Bucharest, Splaiul Independenţei 313, Bucharest, Romania.
[3]Department of Computers and Information technologies, Faculty of Engineering in Foreign Languages, University Politehnica of Bucharest, Splaiul Independenţei 313, Bucharest, Romania.
[4]Doctoral School of Entrepreneurship, Engineering and Business Management, University Politehnica of Bucharest, Splaiul Independenţei 313, Bucharest, Romania.

E-mail: dcrina07@gmail.com

Abstract. Over time, the evolution of technology has left its mark on our lives and has highlighted its need for and importance of all areas of activity, both social and economic. There is a global increase in the trend of automation of work processes and the use of software products more and more often in any type of activity to optimize and increase work capacity. The growth of e-commerce is one of the segments favored by the emergence and evolution of technology, allowing us to always be connected to market demand and supply. In order to improve e-commerce, we need to make sure that the software available to users is adjusted to their needs. In this paper, we propose a technique for analyzing the necessary criteria to develop an automatic testing system for web and mobile applications based on the analysis of user behavior and reports extracted from recorded online traffic. By trying to reproduce the real behavior of users in the testing process it can enhance the risk management and improve the quality of the delivered product.

1. Introduction

Software systems have become key elements in the economic and social activity of our days. The development of a product or a software system goes through many stages before the final product is given to users. These stages are part of the software development process. Simultaneously, with the effective development, the testing process takes place, which is considered one of the most difficult and expensive activities. Testing refers to the validation and verification of a software system and, according to the ANSI/IEEE 1059 standard, it can be defined as a "process of analyzing a software element to detect differences between existing conditions and requirements and to evaluate the characteristics of the software element" [1]. As more and more organizations became aware of the need to deliver quality products and services, the testing process came into their attention and various methods were developed to improve the process. A quality software product means a maintenance or even an increase in the number of users of the product which leads to the success of the business.

IOP Conf. Series: Materials Science and Engineering 1268 (2022) 012015 doi:10.1088/1757-899X/1268/1/012015

Among the most popular models for improving the development and testing process of software products, we can mention CMM (Capability Maturity Model) and TMM (Testability Maturity Model) [2]. These two models are the basis for the development of many other methods by which the most optimal level is desired in terms of testing.

The present article will detail the types of testing currently used in the software field, and then briefly present the business on which the recorded online traffic analyzes will be made. The traffic analysis will be presented in the fourth part of the article, after which some conclusions will be drawn.

2. Overview of the testing process

Testing is a complex process made up of several elements such as: test stages (test planning, test monitoring and control, test analysis, test design, test implementation, test execution, test completion), test levels (component testing, integration testing, system testing, acceptance testing), types of testing (functional testing, non-functional testing, structural testing, change related testing) and testing methods (the most popular methods being Static and Dynamic testing) [3].

Since the article aims to review several criteria for the analysis of online traffic encountered in web and mobile applications to model an automatic testing system, it is necessary to analyze these two testing methods a little more closely.

2.1 Static Testing

"Static testing is testing by examining products (such as requirements specifications, manuals, or source code)" [4]. This type of testing is performed without executing the work product, code, or other testing tool. What defines static testing are reviews or static analysis.

One of the biggest advantages of static testing is the prevention of defects. Static testing takes place before compilation and helps to identify defects in time. Fixing defects found early is less expensive than a fix for a product that is already implemented or worse, which is actively used. Since discussions and meetings between team members are necessary for this type of testing, we can mention that static testing even helps to improve communication between team members. Also, there are certain defects that can only be identified through static testing, not through dynamic testing [3].

2.2 Dynamic Testing

"Dynamic testing is testing by executing the test object, by running an application" [4]. In dynamic testing, the emphasis is placed on behavior visible from the outside and on the way a software product is executed.

In contrast to static testing, dynamic testing consists of finding and fixing defects, therefore, it is more expensive. Dynamic testing is supported by test tools and test cases and involves less communication between team members. Dynamic testing, in turn, incorporates two very well-known methods among testing teams, such as: white-box and black-box:

- White-box: check the flow of input and output data by testing the internal structure, designing and coding the software. The name "white-box" indicates that the tester has access to the code he executes and can understand the actions that led to the output data set.
- Black-box: this testing method emphasizes the inputs and outputs of software applications based entirely on requirements and specifications because the internal code structure and implementation details are unknown.

The testing activity can be done manually or automatically. Manual testing is performed by human resources. This testing is performed more slowly than the automatic testing, but there is a possibility that during the manual testing some defects that an automatic would not be able to detect can be identified. With the apparition of larger and more complex software systems, the idea of automating the testing process has become more popular. Due to the complexity of software systems, the automation of test cases requires great programming skills, and as a result, an automation project can even be seen as a software development project.

The tools that automate the execution of tests are called test tools and they can vary depending on the level of testing at which they are used, for example from the level of unit tests to the level of feature verification tests in Agile projects or to the level of system tests [5].

The most common actions that these test tools, regardless of their nature, can do are: transmitting input data sets (inputs) and other events to the tested product and capturing product responses (output data sets). The development of automated scripts can be adapted for both web and mobile applications. If we are to mention some of the available web tests, we can list: functionality testing, functional testing of web services, regression testing, compatibility testing and so on. [6]. Usually, a web software application used in a business environment is prone to constant changes and it is more feasible to perform manual testing than automatic testing, the test cases needing to be constantly adapted, which implies higher costs and more time. However, it is recommended to create automatic test cases for web applications to ensure that any web product works correctly, which would offer the possibility to reuse and expand on various browsers, platforms, servers and help the identification of problems related to the access time of web pages by users.

Organizations that decide to create a set of automated tests for the web applications they develop must be careful about the test tools they use. One of the important aspects that must be taken into account when choosing a test instrument is the compatibility between the design and implementation of the application and the instrument. The financial side and its performance must also be analyzed. At the moment, the market has a lot of testing tools such as: Selenium, Watir, HP-QTP, LoadRunner, TestNG, WinRunner.

The advantages of automatic testing applied to mobile applications are similar to those mentioned above in the case of web applications. Automated test cases help reduce costs, increase application accuracy, and help identify the causes of defects faster than manual testing.

Before choosing a testing tool in which to develop automated tests for mobile applications, one must consider that the tool supports several programming languages, offers the possibility to run the tests on several operating systems and integrates with the application. Currently there are testing tools for mobile devices such as: TestComplete, Appium, Robot Framework, Xamarin.UITest, Expresso, etc. [7]

To model an automated application testing system, existing studies show that a technical test analyst has to go through a series of tasks to establish some of the system's criteria.

Before the development of an automatic testing system, it is preferable to establish some development stages and to answer some questions such as: which test cases to automate, which is the purpose of our system, do we want to enable a lot of tests to be executed in a short time frame. For the first question we can think about repetitive tests that run from multiple build, tests that tend to cause human error, manual tests that take too much time and tests to run under certain conditions predetermined in the development source code. [3] This way it will be known what type of test it is, for example: regression tests, smoke tests, functional or non-functional testing.

In his book, Chris Goward exemplifies an approach to online traffic analysis and presents a method for prioritizing test cases using web analytics data. This method, called the PIE Method, can also be applied in the identification of test cases that need to be automated. PIE method approaches three factors: potential, importance and ease.

The question addressed when it comes to potential is "how much improvement can be mode on the pages?" [8]. When creating the tests, we must take into account the fact that it will not be possible to test all the pages at the same time, thus, analyzing data with a view to the performance of the application pages, it will be possible to create tests that focus more on the pages with a low performance.

The importance is determined by the question "how valuable is the traffic to the pages?". [8] The most important pages are those that register the highest volume and the most expensive traffic. An important resource for determining the importance of pages is web analytics data. Performance and importance are closely related because there may be cases where pages with very low performance will be identified, but with insignificant importance and thus these tests will not represent a priority.

The last question, "how complicated will the test be to implement?", establishes the degree of difficulty required to carry out the test. The purpose of developing an automatic testing system is to use less time and fewer resources to record the same or a higher efficiency.

3. Business flow presentation

The present article wants to answer some necessary questions to identify the criteria for analyzing the online traffic encountered in web and mobile applications with the aim of designing an automatic testing system. But before proceeding to the traffic analysis and setting the objectives of this automatic system, it is necessary to know the business flow.

The application for which we will later analyze the online traffic is addressed to e-commerce consumers in Europe by providing greeting cards and birthday gifts. In short, when the application is accessed both web and mobile, it will open a main page, home page, where you can find information and images of the products that can be purchased. In the footer of this page there is administrative information. From the home page, a user can access several parts of the site: either go to the log in or create an account section, then start the flow to place an order by choosing a product or access the page of product filters to search for the most suitable gift.

To place an order, the user can choose to create an account with a username and password where they can later save their contacts and special events, save certain cards to send them later, keep a history of orders and access the newsletter section, or place an order as a visitor.

Once the filters page has been accessed, the user has at their disposal several options that they can select so that it is easier to find the desired card or gift (Figure1).

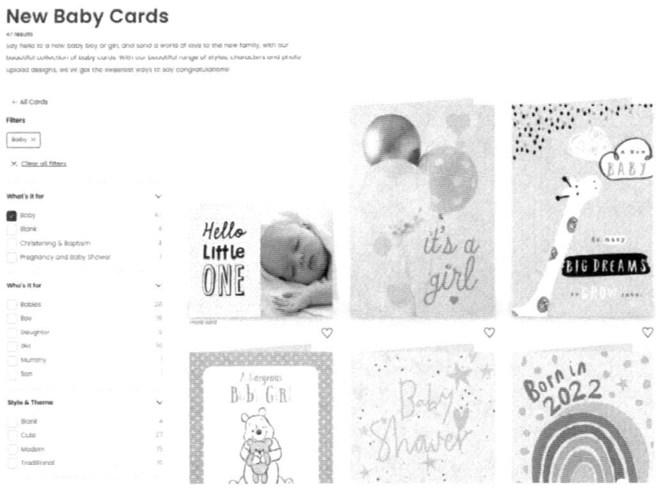

Figure 1. List page of products

Once the card has been chosen, a page with more details about the product will open. The user can go further with the desired card and access the "size picker" page to opt for a standard card size or a larger one.

After all these criteria have been chosen, the application allows the user to personalize their card and write a message to the person to whom he will send the gift. Usually, this page is of great interest and its design can be observed in Figure 2.

Until the user completes and pays for an order, a page follows where they are recommended to associate the chosen card with a gift. Afterwards, the recipients and the delivery date of the chosen card/gift will be chosen, and the order will be paid for.

The business flow of the mobile application is not much different from that of the web application. The pages, products, and steps to follow to place an order are identical, the differences consist more in the Front-End elements.

IOP Conf. Series: Materials Science and Engineering 1268 (2022) 012015 doi:10.1088/1757-899X/1268/1/012015

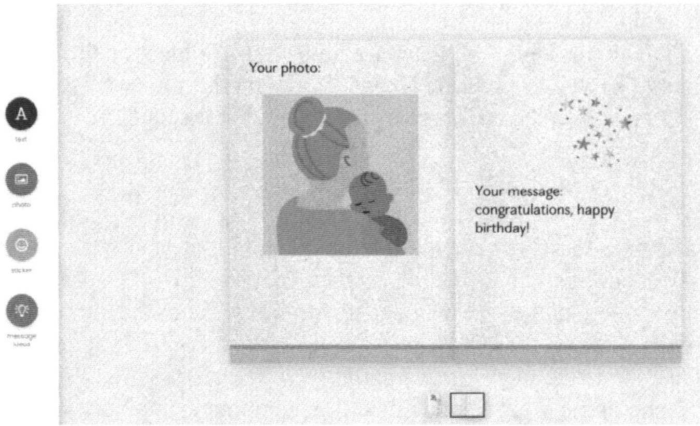

Figure 2. Card customization page

4. Traffic analysis

Data traffic analysis will be done with the help of reports extracted from Google Analytics. "Google Analytics is a platform that collects data from your websites and apps to create reports that provide insights into your business." [9]

The study is carried out on the real application described in the previous chapter over a period of one week, in the time interval August 14 - August 20, 2022, using the traffic recorded in the Netherlands. The data in the following tables, except those related to the average time spent on each page, are multiplied by a coefficient between 0 and 1 in order to preserve the confidentiality of the data but to emphasize the trend encountered in the behavior of users.

The process of automating test cases is a complex one, and even more complex than developing test cases is their maintenance. It is possible that at some point, for an automation process developed for applications or parts of applications that change their content constantly, it will be more difficult to update the tests and to identify the manual testing as being more feasible than the automatic one.

To decide which of the test cases deserve to be automated, one can review most of the test cases for manual testing and create a checklist with those that deserve to be executed either manually, automatically or by using a combination of manual and automatic testing depending on the strategy used (as it can be observed in Figure 3).

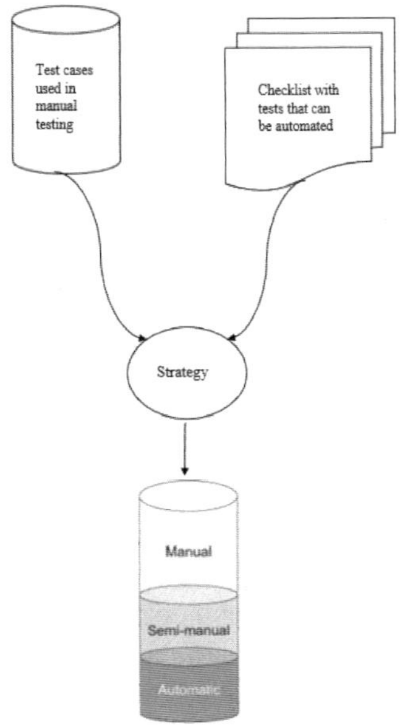

Figure 3. Differentiation between test cases

For the development of automatic test cases, we can also use the information regarding the behavior of the users on the application in question using the reports extracted from Google Analytics as mentioned above.

For example, in the first stage, we can check where more traffic is registered, on a web application or on a mobile one.

Table 1. Number of accesses via device category

Device Category	Number of accesses via device category
mobile	56486
desktop	24599
tablet	3116

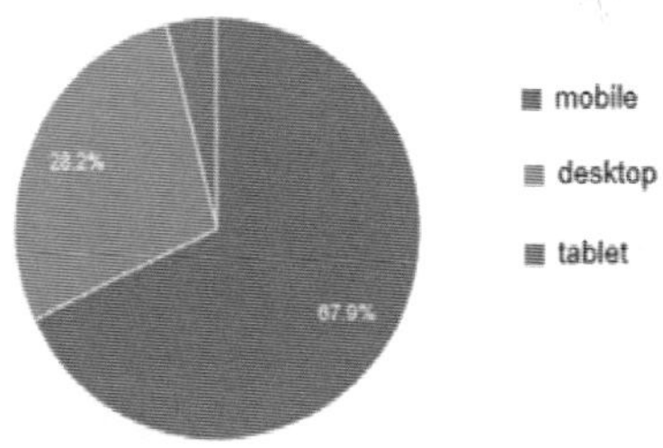

Figure 4. Device Category

For each of these types of devices, the tool can identify the browser used by users and the operating system. This information is useful in two ways - to identify the appropriate automatic testing tool and to develop efficiently tests that will run the application through those technologies. In the present case, the statistics can be seen in figure 5.

IOP Conf. Series: Materials Science and Engineering 1268 (2022) 012015 doi:10.1088/1757-899X/1268/1/012015

Figure 5. OS & Browsers used for accessing the cards website

More interesting than that is that using these types of reports, test flows can be created to imitate user behavior by identifying the most liked pages and products among buyers. By developing automatic tests close to user behavior, we can support and help project risk management because running these tests will ensure the correct functionality of the application and eliminate the risk that a customer will not be able to complete their order if they are in the range defined as being the most used. It should be mentioned that the "/" page in the figure 6 refers to the home page of the application.

Table 2. Top 10 most accessed pages

Page	Pageviews (1,071,831)
/	**75,108** (7.01%)
/kaarten/customizedcard	**73,446** (6.85%)
/kaarten/chooserecipients	**44,947** (4.19%)
/kaarten/shoppingcart	**44,088** (4.11%)
/kaarten/chooseordercardtype	**41,466** (3.87%)
/kaarten/choosedeliverydate	**41,320** (3.86%)
/kaarten/choosegiftcategory	**39,735** (3.71%)
/kaarten/adyenpayment	**37,392** (3.49%)
/wenskaarten/	**22,201** (2.07%)
/l/verjaardagskaarten/	**20,879** (1.95%)

Table 3. Average time spent on each page

Page	Avg. Time on Page
/	0:00:29
/kaarten/customizedcard	0:04:32
/kaarten/chooserecipients	0:01:19
/kaarten/shoppingcart	0:00:30
/kaarten/chooseordercardtype	0:00:12
/kaarten/choosedeliverydate	0:00:18
/kaarten/choosegiftcategory	0:00:11
/kaarten/adyenpayment	0:01:07
/wenskaarten/	0:00:21
/l/verjaardagskaarten/	0:00:45

Although the top shows that the filter page is the most accessed by users, it is important to look also at the time spent by customers on these pages. The page on which the user spends the most time is the one where they edit their chosen card. This is the reason why it is necessary that the implementation of some test sets verifies as many functionalities of this page as possible:

After the statistical analysis made in the pages above for the site with cards, the regression represents some of the most important tests to be automated because it helps to identify possible errors encountered after a software program has been modified. That's why, according to the data extracted above, an automated regression testing system is required to check the most frequently accessed pages, such as the filter page or the personalization of the card. In order to achieve a complete flow and as close as possible to the user's behavior, it will be necessary for the automatic regression tests to run our software in mobile and desktop mode using both iOS and Android operating systems. Regarding the browsers used, we will focus on the two: Safari and Chrome.

During the flow of placing an order, every time one of the pages of the application does not work as expected, and the user cannot go to the next page, the automatic regression test system will display an error and a screenshot of the page where the application is not compliant so that it is easy to identify the problem.

IOP Conf. Series: Materials Science and Engineering 1268 (2022) 012015 doi:10.1088/1757-899X/1268/1/012015

Since we notice that the page where personalization of the cards is allowed is the second most frequently accessed page with an average time spent on the page of approximately 4 minutes, it will be necessary to create some exploratory tests to make sure that the system works perfectly and the user can personalize his card by attaching images, customizable text or even by using stickers. These types of testing are known because they check if problems fixed in the past reappear in newer versions. For this page, it is important to check this aspect because any problem related to this section of the application directly impacts the interaction between the application and the user.

5. Conclusion

From the extracted graphs, we can see the tendency of users to access software products using mobile devices rather than desktop ones. However, whether it is a web application or a mobile application, the quality of the software product must be as high as possible. Test team members must be aware of process approaches, test tools, and specific checks. Once the criteria of an automatic testing system are well defined, they can bring important benefits such as: reducing repetitive manual work, widening the coverage rate, reducing costs, quick access to test results and facilitating the identification of the source of defects.

When we decide that we want to automate the testing of a software product or part of its testing, we must not forget that in addition to the advantages mentioned above, there are also potential risks such as: underestimating the effort to maintain active tests, limitations of the testing tool used, incompatibility of the test tool with the platforms or technologies/operating systems used by the users, the need for personnel who acquire the necessary knowledge to implement the test cases.

In conclusion, the automation of test cases is a process of increasing the quality of software products for which a pre-stage of analysis and correct identification of the elements that can be automated is necessary.

References

[1] IEEE Standards Board 1994 *IEEE guide for software verification and validation plans*. Institute of Electrical and Electronics Engineers

[2] Paulk MC, Curtis B, Chrissis MB, Weber C v. 1993. *Capability Maturity Model SM for Software, Version 1.1.*

[3] Bath G and McKay J *The Software Test Engineer's Handbook -2nd Edition* Santa Barbara California Rocky Nook

[4] Marselis R and Veenendaal B and Geurts D and Ruigrok W 2022 *Quality for DevOps* Vianen Netherlands Sogeti

[5] Black R 2015 *Advanced Software Testing* – Vol 1 Santa Barbara California Rocky Nook

[6] Sharma M and Angmo Rigzin 2014 *Web Based Automation Testing and Tools IJCSIT* **5** 908-912

[7] Cannizzo F and Marcionette G and Moser P 2008 *Evolution of the Tools and Practices of a Large Distributed Agile Team* Proceedings of the Agile 2008 Conference, IEEE Computer Society 513-518

[8] Goward C 2012 *You Should Test That! Conversion Optimization for More Leads, Sales and Profit or The Art and Scrience of Optimized Marketing* Canada Sybex

[9] *How Google Analytics works - Analytics Help* [Internet]. [cited 2022 Sep 30]. https://support.google.com/analytics/answer/12159447?hl=en

A functional design analysis of a robotic-guided brachytherapy instrument

Paul Tucan[1], Florin Graur[2], Andrei Caprariu[1], Corina Radu[2], Alin Burz[1], Calin Vaida[1], Nadim Al Hajjar[2] , Tiberiu Antal [1] and Doina Pisla[1,*]

[1]CESTER, Technical University of Cluj-Napoca, 400114 Cluj-Napoca, Romania

[2] Iuliu Hatieganu University of Medicine and Pharmacy of Cluj-Napoca, 400012 Cluj-Napoca, Romania

*Corresponding author: doina.pisla@mep.utcluj.ro

Abstract. The paper presents the functional design analysis-based improvements of a robotic guided medical instrument for brachytherapy. The initial design of the medical instrument is analyzed, and its weaknesses are identified and analyzed from several points of view (functionality, ergonomics and materials). The design analysis reveals the specific improvements for the instrument components to achieve a better functionality of the medical instrument and increase the accuracy and safety of the robotic-assisted medical procedure.

Keywords. Design analysis, robotic-guided brachytherapy, FEA, parallel robot.

1. Introduction

Liver cancer is among the most diagnosed cancers around the world and the third cause of death [1]. Hepatocellular carcinoma (HCC) is mainly caused by infection with hepatitis B and C viruses, but it can be also caused by heavy use of alcohol, morbid obesity, diabetes, and ingestion of certain aflatoxins (poisonous carcinogens produced by mold) [2]. Standard care for HCC implies surgical resection and transplantation, but these options are not always available for the patient and in most cases (circa 80%) the tumors are inoperable and/or unresectable [3]. To overcome the predicament of inoperable tumors some locoregional therapies have been developed in time such as radiofrequency ablation, arterial chemoembolization and brachytherapy [4]. Brachytherapy implies the positioning of radioactive seeds with high dosage and limited spread, straight within the tumor with the target of causing minimal damage to the healthy tissue [5]. The success of the brachytherapy procedure is highly dependent on how accurately the brachytherapy needles are inserted within the tumor. Due to their improved accuracy and stability during the procedure, robotic devices begun to be used in the brachytherapy procedure.

Lin [6] presents an interventional 5 DoF robotic system and a needle actuator used for inserting the brachytherapy needle. The experimental test performed with the instrument revealed a standard deviation of the tissue targeting error of 0.33 mm and an accuracy of needle placing of 1.10 mm.

Yu [7] presents a 6 DoF platform than can position and orient the therapy needle in 3D space. The system is used for prostate brachytherapy. The system was clinically tested using 10 mm/sec speeds for the positioning of the needle and 50 mm/sec speed for inserting the needle. The mean error at the tip of the needle recorded after 100 repetitive insertions was ±0.2 mm with a standard deviation of 0.045 mm.

Strassmann [8] tested the accuracy of robotic assisted needle insertion with the one of the conventional template-guided using a fresh porcine abdomen. The result revealed a better accuracy in the case of robotic procedure 1.8±0.6 mm using a insertion velocity of 2.7mm/sec while the template-guided procedure revealed an inferior accuracy of 2.7 ±0.7 mm. When higher speeds were used for the robotic insertion, there were no significant differences recorded.

This paper presents a functional design analysis of a medical instrument designed for robotic assisted brachytherapy using a parallel or a serial robotic arm for positioning and orienting the medical instrument. Starting from the initial design of the medical instrument, a redesign process is carried on in order to improve the functionality of the medical instrument and at the same time the safety of the robotic assisted procedure. The requirements of the instrument were established by medical specialists based on their experience. The objective of the research is to achieve a more compact and robust design of the brachytherapy instrument that can easily operate multiple needle insertions under a CT Scanner with high accuracy.

The second section contains information regarding the current state of the medical instrument for brachytherapy, system integration and functionality. The third section contains functionality issues identified during the functional design analysis followed by redesign of the instrument in the fourth section, conclusions, and acknowledgment.

2. Initial design of the brachytherapy instrument

The brachytherapy instrument was designed to work together with a 5 DoF parallel robot called PARA-BRACHYROB, developed in CESTER, Technical University of Cluj-Napoca [9]. The instrument is attached to the robotic system using two universal joints (Figure 1). The robot places the brachytherapy instrument based on a mathematical model which aligns the needle with the linear insertion trajectory and positions the tip of the needle close to the patient skin. In the second procedural step, the robot stops, and the instrument performs the linear insertion of the needle with real-time force monitoring [10]. When the target point is reached, the needle is released, the instrument is retracted, and the next needle is loaded from the needle storage area. The process is repeated until all the needles are inserted (as requested by the medical experts). The motions of the robotic system are sequential, while the robot moves the instrument is in standstill and vice versa. Single use needles are used for the entire medical procedure. The robot and the instrument are also sterilized before each procedure starts.

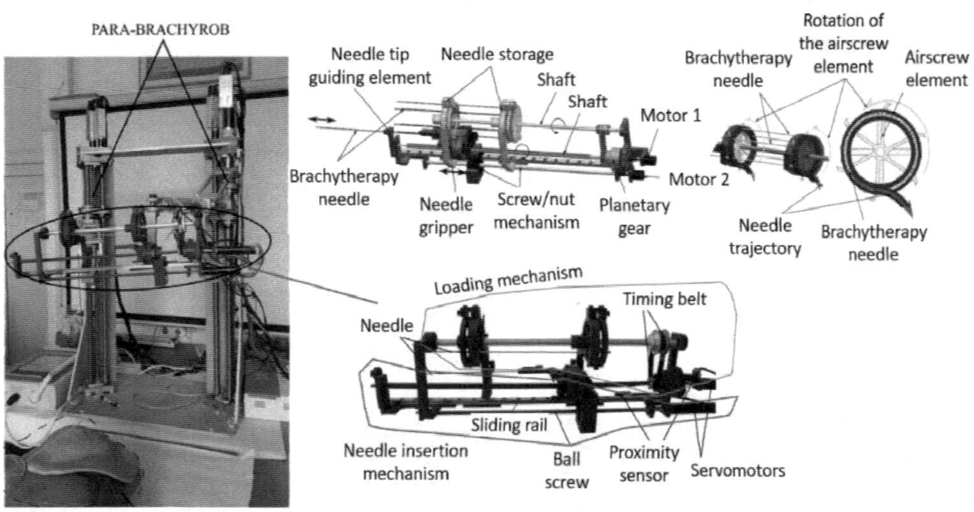

Figure 1. Brachytherapy instrument integrated within the PARA-BRACHYROB

The initial instrument prototype is composed of two modules: the needle loading mechanism and the needle insertion mechanism (Figure 1). The loading mechanism positions needles in the insertion device, one by one from the needle storage which can hold up to 6 needles. The insertion mechanism performs the insertion of the loaded needle on a linear path until the desired depth is reached. Each module is actuated by an individual motor. The two modules are working together in a synchronized manner. Figure 1b presents the most important components of the instrument [10]. The actuators used have a 0.2 Nm output torque. Reduced friction coefficients were employed in the rail/sledge mechanism with sliding rails and ball screw/nut mechanism. A professional 3D printer was used to 3D print the remaining parts of the medical device using ABS plastic material.

The detailed working principle of the instrument is described by presenting the three working phases of a complete functioning cycle all correlated with the positioning and orienting motion of the robot:

A. **Needle Loading Phase**: Motor 1 actuates the rotation of the needle storage airscrew element and the needle loading gripper. As shown in Figure 1, the needle is guided by the storage airscrew element through a circular channel. At the end of the channel, it is cached by the jaws of the gripper. The two shafts of the needle loading module are actuated simultaneously by the same motor. The motion transfer mechanism is composed of a planetary gearbox with a rotation ratio of 1:6 and a belt transmission which is generating the desired output motion: the two shafts are rotating in opposite directions and by a complete rotation of the gripper's shaft, the shaft of the airscrew element is performing a 60 degrees rotation. The needles are placed inside the repository on a circular pattern at a 60-degree angle. The described layout allows the needles to be loaded from the repository one by one, in a controlled manner. The end of the loading phase consists of the fixation of the loaded needle between the two rotary jaws of the gripper.

B. **Needle Insertion Phase**: by actuating motor 2, through a lead screw mechanism, the needle is moved on a linear path until the target coordinates at the tip of the needle are reached. The insertion phase ends with the needle releasement by actuating motor 1 at 30 degrees in the opposite direction of the loading.

C. **Homing and reloading**: after the insertion of the needle is completed, both needle insertion and needle loading modules are sent to their home positions. In order to release the needle inside the tissue without inflicting damage, the robot must retract the instrument on the same linear path that was used for the needle insertion until the exterior cannula of the needle is outside the patient.

3. Weaknesses identification

By collecting data from initial experiments and by visual inspection, the weaknesses of the instrument have been identified and classified.

Design weakness: this category encompasses some features that were not present or not properly designed in the initial state of the instrument. The reaming of the needle tip guiding element is too large compared to the diameter of the needle. In this case, it is not able to offer adequate radial support during the insertion and there is a higher possibility that the needle could bend and eventually deflect from the desired trajectory during the insertion process.

Mechanical strength weaknesses: (1) One of the transmission mechanism's shafts is not rigid enough. It was manufactured through 3D printing using ABS. The shaft can easily bend, affecting the overall accuracy of the instrument. (2) Some other 3D printed components are breaking or excessively bending. (3) The instrument was designed for being mounted on a parallel robot - PARA-BRACHYROB [9]. The mounting interface is also printed using ABS. The two arms that should support the entire weight of the instrument can easily bend. (4) The circular needle repository showed elastic deformations during tests failing to load properly the needle in 30% of the instances.

Ergonomic weaknesses: The overall size of the instrument can be a challenge for operating the instrument, especially for multiple needles insertion when the insertion of a new needle must not affect the already inserted ones. The most considerable part of the instrument's volume is occupied by the needle loading mechanism.

4. Redesigning of the brachytherapy instrument

The first challenge in the redesign process is the ergonomic one. The overall dimensions of the entire instrument had to be reduced to achieve a more compact one that can be better manipulated. As mentioned, the needle loading module is the largest one. In this case, redesigning the needle loading module from the sketch is the first step of the process.

Figure 2 shows the new instrument, comparing its size with the size of the original instrument. The overall volume of the redesigned instrument was significantly reduced. Around 36 mm were reduced in the X direction, 92 mm from the Y direction, and 49 mm from the Z direction. The volume of the surrounding box of the original instrument in the home position was 9.64 liters. For the redesigned instrument, the volume of the surrounding box is 2.54 liters.

Figure 3 presents the components of the redesigned needle loading module in the initial state. The six needles are placed inside the needle repository. The needle repository is made of two components: the fixed, upper (1) repository, and the mobile, lower repository (2). The gripper is composed of two jaws. The lower one is the solenoid-driven jaw (3) and the upper one is the spring-driven jaw (4). The solenoid-driven jaw is actuated by the solenoid (5).

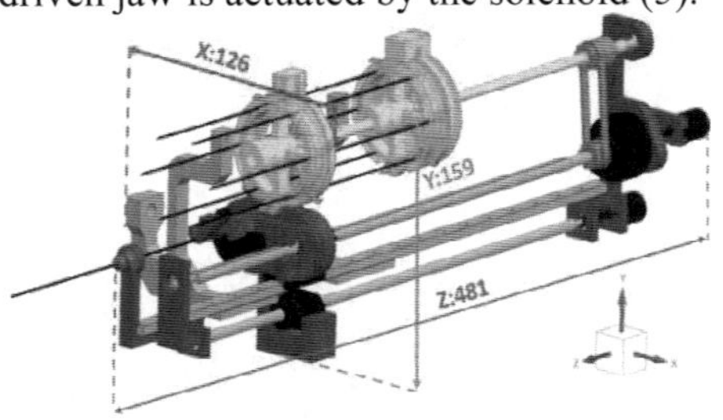

a. Initial instrument

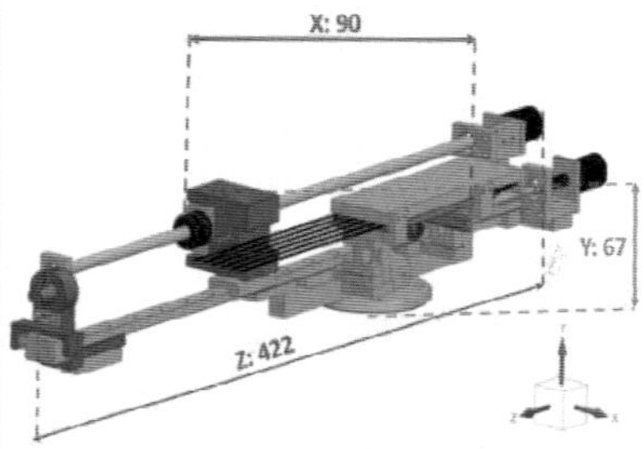

b. Redesigned instrument

Figure 2. Initial instrument vs redesigned instrument. Dimensional comparison

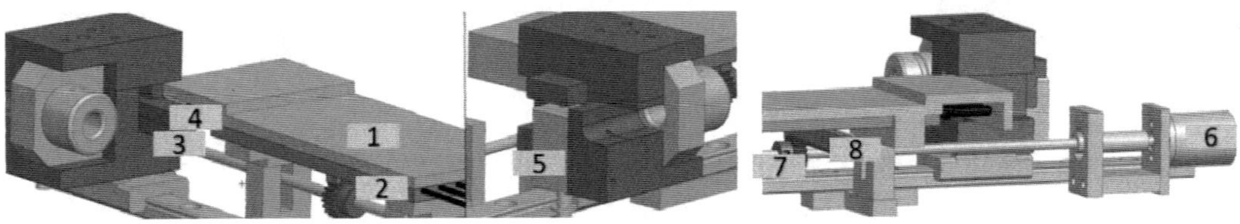

Figure 3. Redesigned needle loading module

In the initial state (home position) the needles are aligned with the gripper in a common plane. For loading a needle, the mobile storage (2) is actuated by the motor (6) through the rack (7)-pinion (8) mechanism. Because of its geometry, the spring-driven jaw is lifted by the needle's head while the mobile repository is translated. In the next phase, the motor-driven jaw (3) is actuated, and the needle is lifted out of the repository while being pressed by the two jaws. After the needle is lifted out of the repository, the mobile repository moves back to the home position. The same cycle is repeated for loading the other 5 needles.

As identified before, the inappropriate radial support of the needle was one of the weaknesses of the original instrument. The needle guiding module was designed with the purpose of supporting the needle during the insertion process. The needle guiding module is presented in Figure 4.

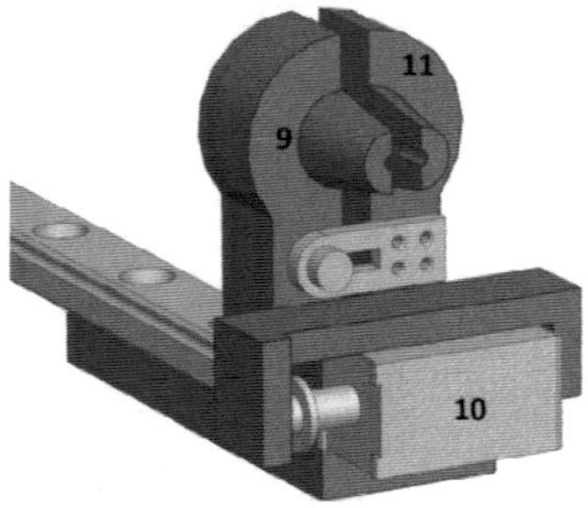

Figure 4. Needle guiding module

The module is composed of two jaws: the mobile jaw (9) which can be actuated by the solenoid (10) and the fixed jaw (11). After the tip of the needle has passed through the two jaws of the guiding module, the jaws are closing and guide the needle in a linear trajectory, offering the necessary redial support.

A new mounting adaptor has been designed for the instrument. The adaptor allows the instrument to be mounted on a medical collaborative robot- KUKA LBR IIWA 7 R800 [11] and, by using an adapter flange, the instrument can be also mounted on the parallel robot – PARA-BRACHYROB. The new mounting adaptor is presented in Figure 5a. The blue component from Figure 5a is the support bar which is mounted on the rail using bolts. The component highlighted using green color is the flange. The flange is mounted on the support bar. The center of gravity of the flange is coincident with the projection of the center of gravity of the instrument.

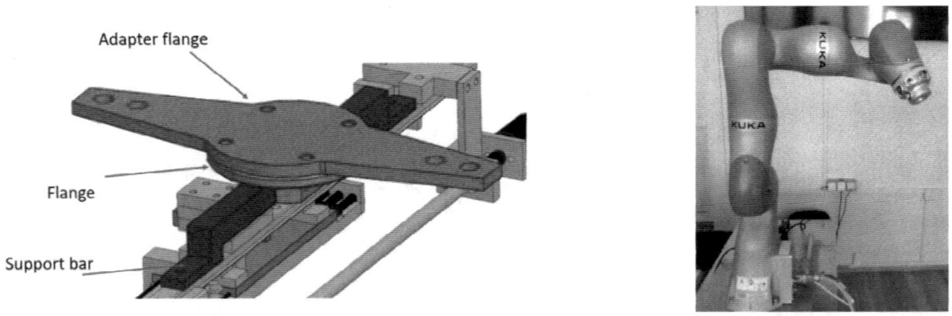

a. Mounting adaptor b. Kuka iiwa robot

Figure 5. Redesigned mounting adaptor (a) and Kuka iiwa robot (b)

5. Design analysis

The general mechanical strength of the instrument, besides redesign, is also improved by manufacturing the structural components out of steel and aluminum. In order to test the strength of the redesigned instrument, several static and quasi-static finite element analyses have been performed. The instrument was modeled using solid elements (hexa first order or second order pyramids) with a mesh size of around 0.5 mm for the most critical areas or larger for the rest of the areas.

The first CAE simulation tests the strength of the mounting interface of the instrument. The simulation only contains the mounting interface, the support bar, and the rail. The mass of the other elements is simulated using the mass element. A boundary condition is used to fix the mounting interface on the 4 holes. For the presented model, four sinusoidal 10g shocks were simulated, one in each +X,

+Y, -Y, and +Z direction. The duration of each shock step was 6ms. Each shock was followed by a relaxation step of 12ms.

The highest von Mises stress of the Flange appears during the +X shock. The stress is around 78 MPa, lower than the yield limit of the material. Additionally, during the X shock simulation, the highest von Misses value can be noticed for the support bar. No plastic deformations will appear since the stresses are under the yield limit of the material. Figure 6 a shows the maximum von Mises stress during the +X shock. The highest von Mises stress value for the rail appears during the +Y shock simulation. The value of the stress is still under the yield limit of the material. Figure 6 b presents the maximum stress state of the -Y shock.

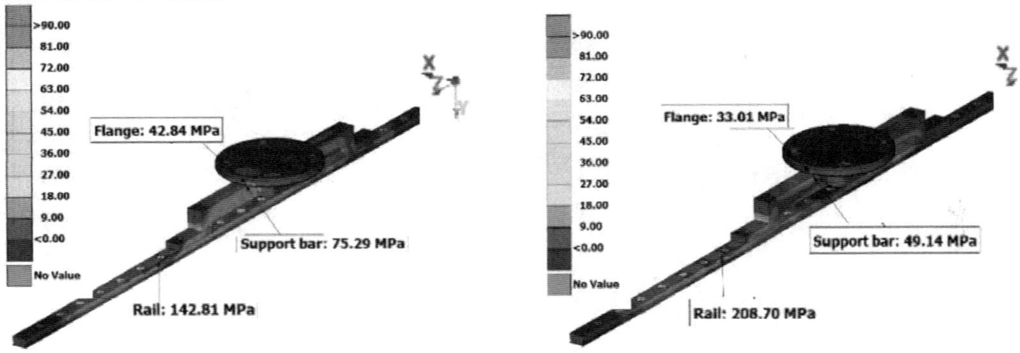

a. Shock X. Simulation results b. Shock Y. Simulation results

Figure 6. Mounting adapter simulation results

For testing the strength of the structural components of the redesigned instrument, the components that should intuitively be the most sensible in terms of mechanical resistance are tested in the following simulation. Figure 7. presents the load case scenario of the simulation

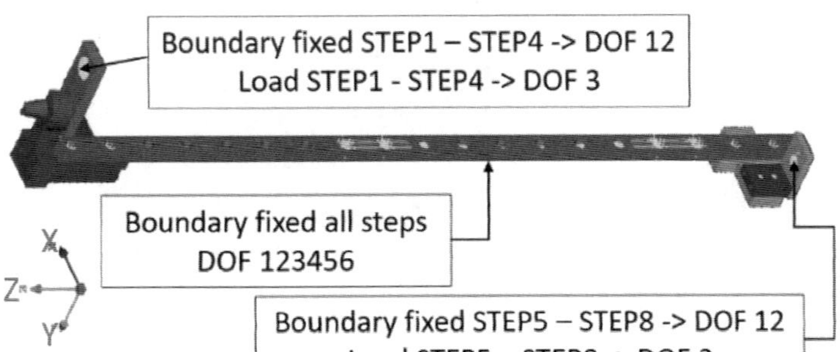

Figure 7. Sensible structural components simulation – loadcase scenario

The simulation consists of eight static steps. In the first step, a concentrated load (magnitude = 15 N, direction = +Z) is applied to the kinematic coupling from the front of the instrument, as pointed out in Figure 7. In the second step, the load is removed. In the third step, another concentrated load is applied to the same node (magnitude = 15 N, direction = -Z). In the fourth step, the load is removed. In the fifth step, another concentrated load (magnitude = 15 N, direction = Z) is applied to the kinematic coupling from the back of the instrument, as pointed out in Figure 7). In the sixth step, the load is removed. In the seventh step, another load is applied on the same node (magnitude = 15 N, direction = -Z). In the last step, the load is removed.

Figure 8.a presents the maximum von Mises stresses of the front structural components achieved during the entire simulation. The stresses are reached at the end of the third step – load in -Z direction.

The area where the highest stresses appeared is on the contact between the bearing support component and its base. The result of the simulation shows that the maximum von Mises stresses are under the yield strength limit of the material (90 MPa). Figure 8 b presents the maximum von Mises stresses of the back structural components achieved during the entire simulation. The stresses were reached at the end of the fifth step – load in +Z direction. The area where the highest stresses appeared is on the contact between the motor support component and its base. The result of the simulation shows that the maximum von Mises stresses are under the yield strength limit of the material (90 MPa).

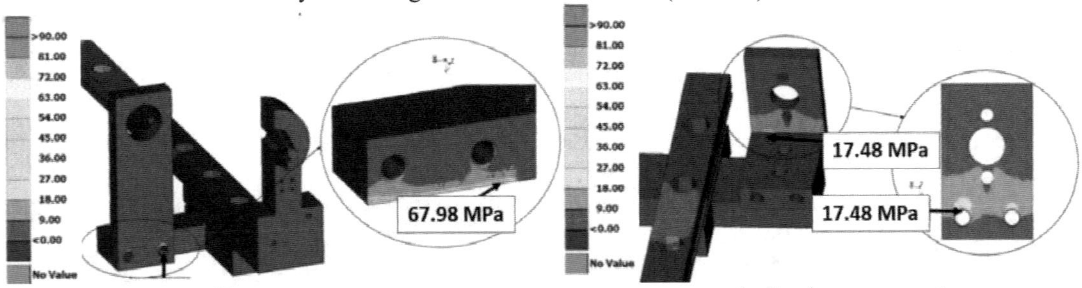

a. Front components b. Back components

Figure 8. Sensible structural components. Simulation results

Next, a gripper opening simulation is carried on to simulate the strength of the gripper's jaws, by simulating the opening and closing of the gripper while the translation of the jaw is blocked. Figure 9 presents the loadcase scenario of the simulation. During the entire simulation, the spring jaw (gray component) is fixed on the upper surface and the sides of both jaws will only be free to translate in the Y direction. In the first step, a concentrated load will be applied on the motor jaw (red component). In the second step, an additional fixed boundary condition will be applied on the bottom surface of the motor jaw and the load will be applied in the -Y direction.

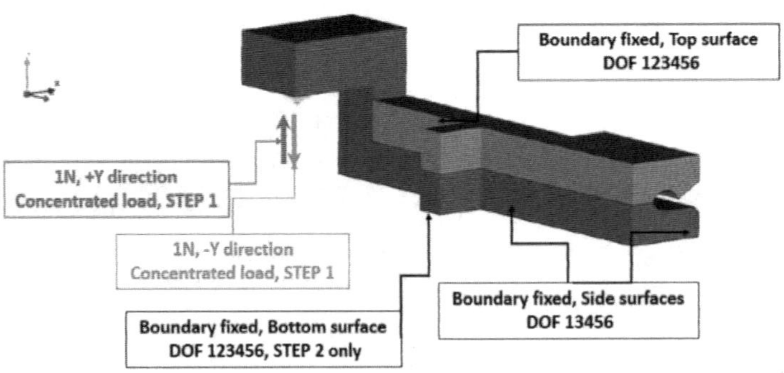

Figure 9. Gripper opening simulation. Loadcase scenario

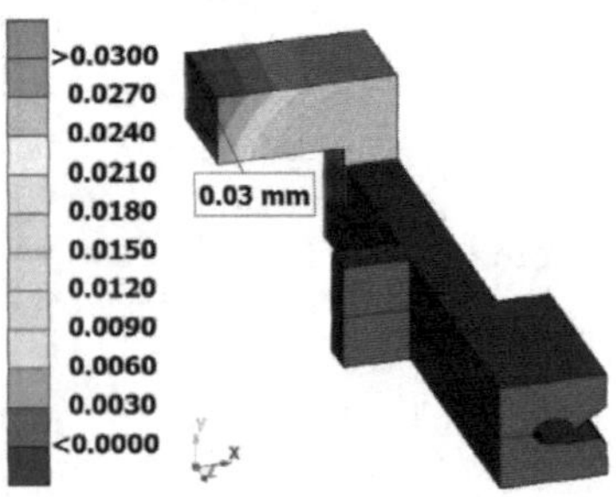

Figure 10. Gripper opening. Simulation results.

Figure 10 presents the maximum total deformation of the gripper opening simulation. The stresses and deformations in the gripper opening simulation are very small.

6. Conclusion

The paper presents a functional design analysis of a robotic instrument resulting in a series of improvement solutions. The design, ergonomic, and mechanical strength weaknesses of the initial prototype of the device have been identified and analysed. The identified weaknesses have been eliminated by redesigning specific components and subassemblies. A series of finite element simulations was performed to test the stiffness of the new instrument. The obtained results revealed a more compact and robust instrument based on the initial one. Further work implies the prototype development of the newly designed medical instrument and the development of the control system of the instrument joined with the control system of the instrument manipulating robot. Furthermore, for the validation of the design, the robotic system will be tested in the laboratory and in the hospital together with the medical experts.

Acknowledgment: This work was supported by a grant of the Romanian Ministry of Research and Innovation, CCCDI – UEFISCDI, project number PN-III-P2-2.1-PED-2019-4375/397/2020 (OnTarget), within PNCDI III and by the project POCU/380/6/13/123927—ANTREDOC, "Entrepreneurial competencies and excellence research in doctoral and postdoctoral studies programs", co-funded from the European Social Fund through the Human Capital Operational Program 2014–2020.

References

[1] Rumgay, H.;Ferlay, J.; de Martel, C., Georges, D.; Ibrahim, A.S.; Zheng, R.;Wei, W.; Lemmens, V.E.P.P; Soerjomataram, I.; (2022) Global, regional and national burden of primary liver cancer by subtype, European Journal of Cancer ,161:108-118.

[2] Chuang, S.C.; Vecchia, C.L.; Boffetta, P.; (2009) Liver cancer:descriptive epidemiology and risk factors other than HBV and HCV infection, Cancer Lett, 286(1):9-14.

[3] Brar, G.; Kesselman, A.; Malhotra, A.; Shah, M.A.; (2021) Redefining Intermediate-Stage HCC Treatment in the Ero of Immune Therapies, JCO Oncology Prcticem 18:35-41.

[4] Heimbach, J.K.; Kulik, L.M.; Finn, R. S.;Sirlin, C.B.; Abecassis , M.M.; Roberts, L.R.; Zhu, A.X.; Murad, M.H.; Marrero, J.A.; (2018) AASLD guidelines for the treatment of hepatocellular carcinoma, Practice Guideline>Hepatology, 67(1):358-380.

[5] Vaida, C; Al Hajar, N.; Lazar, V.A.; Graur, F.; Burz, A.; Elisei, R.; Mois, E.; Pisla, D.; (2019) Robotics in Minimaly Invasive Procedures:History, Current Trends, and Future Chalanges, 6th International Conference on Advancements of Medicine and Health Care through Technology, 17-20 October, 2018 Cluj-Napoca, Romania.

[6] Lin, X.; Zhou, S.; Wen, T.; Jiang, S.; Wang, C.; Chen, J.; (2021) A novel multi-DoF surgical robotic system for brachytherapy on liver tumor: Desing and Control, International Journal of Computer Assisted Radiology and Surgery, 16:1003-1014.

[7] Yu, Y.; Podder, T.K.; Zhang, Y.D.; Ng, W.S.; Misic, V.; Sherman, J.; Fuller, D.; Rubens, D.J.; Strang, J.G.; Brascchio, R.A.; Messing, E.M.;(2007) Robotic system for prostate brachytherapy, Computer Aided Surgery, 12(6):366-375

[8] Strassmann, G.; Olbert, P; Hegele, A.; Richter, D.; Fokas, E.; Timmesfeld, N.; Hofmann, R.; Engenhart-Cabillic, R.; (2011), Strahlentherapie und Onkologie, 187(6):367-372.

[9] Popescu, T;.Kacso, A.C., Pisla, D., Kacso, G.;(2015) Brachytherapy nedt generation:robotic systems, J Contemp Brachytherapy, 7(6):510-514.

[10] Birlescu, I.; Craciun, F.;Vaida, C.; Gherman, B.; Pisla, D.; (2017) An Innovative Automated Instrument for Robotically Assisted Brachytherapy Used in Cancer Treatment, Acta Technica Napocensis, Series: Applied Mathematics, Mechanics and Engineering, 60(4):633-638.

[11] Kuka Ag. http://www.kuka.com (2022)

IOP Conf. Series: Materials Science and Engineering 1268 (2022) 012017 doi:10.1088/1757-899X/1268/1/012017

Development and control of a virtual industrial process for assembling computers

Goran Munđar[1]

[1]Laboratory for Mechatronics, Faculty of mechanical engineering, University of Maribor, Smetanova ulica 17, 2000 Maribor, Slovenia

[1]goran.mundjar@um.si

Abstract. A virtual manufacturing method represents the emerging approach that allows companies to improve their processes in order to introduce new products more quickly to their markets in a cost-effective manner. This involves building a synthetic and integrated environment based on software tools and systems, such as Virtual Reality and Simulation, to facilitate such processes. This paper presents the development of virtual industrial environment, highlighting the development and control of virtual industrial process for assembling computers by combining two simulation software tools. Development of industrial process was performed using Factory I/O software tool, while the control program was developed using FluidSim software tool. This paper also highlights the usage of OPC communication protocol, that was used for communication between software tools during simulation.

1. Introduction

With the great technological development in industry around the world, the knowledge and skills required for the successful production of industrial applications are changing more and more. When planning and managing them, we encounter complex problems that require multidisciplinary knowledge from various fields, such as mathematics, physics, computer science, electrical engineering and logic programming. Many components such as sensors, actuators, electrical, mechanical, pneumatic and hydraulic drives are used in industrial applications. Integrating these components into working industrial applications requires a lot of time, money and research into the operation of systems from different manufacturers [1]. In recent years, virtual systems have been developed to facilitate the planning of industrial applications. Virtual systems represent graphical computer simulations that enable the simulation of dynamic systems and are controlled by peripheral control devices that can be simulated or real [2]. The main purpose of virtual systems is to enable engineers to inspect and evaluate the performance of developed industrial applications via computer, at lower costs and without the risk of injury to persons or machines during development and programming of industrial applications [3].

In this work, with the help of two software tools for simulating virtual systems, we created a more complex industrial process for assembling computers, its control, and simulated this process using a computer simulation. In the first part we will present the creation of an electro-pneumatic industrial process with all components in the "Factory I/O" software tool. "Factory I/O" enables the creation and simulation of a virtual industrial environment using various industrial components from major manufacturers [4]. Created virtual system can be controlled using real PLC controllers or using virtual controllers.

In the second part, we will present the creation of a sequential control chain for controlling the industrial process, created in the first part. For this we will use the software tool "FluidSim". "FluidSim" is a software environment that enables the creation and simulation of electro-pneumatic, electro-hydraulic, digital and electronic systems [5].

In the third part, we will perform a simulation of an industrial process created in "Factory I/O" and controlled by a sequential controller created in "FluidSim". For this we will use the program "Festo Didactic EzOPC" [6], which enables communication between "Factory I/O" and "FluidSim" via the OPC Data Access communication protocol.

1.1. Factory I/O

"Factory I/O" is a software tool that enables the creation and simulation of virtual industrial processes. These processes can consist of components that represent different parts of industrial processes. Processes made in "Factory I/O" are usually controlled by PLC controllers, as they are the most widely used in industry. In addition to PLC controllers, controllers that work with TCP/IP and OPC communication protocols can be used to control processes in "Factory I/O".

"Factory I/O" contains a large library of basic industrial components such as sensors, pneumatic cylinders, electric motors, conveyor belts, boxes, etc. Most of these components have digital or analog inputs and outputs that can be controlled using the aforementioned controllers. In "Factory I/O", some industrial processes are already made and ready to be controlled.

1.2. FluidSim

"FluidSim" is a software tool for creating, simulating, teaching and studying electro-pneumatic, electro-hydraulic, digital and electronic systems. "FluidSim" combines a circuit diagram editor with detailed descriptions of all components, component photos, operation animations and video sequences.

In "FluidSim" it is possible to create control programs and use them to control the created systems within the program itself. In addition, it enables the control of external processes using communication protocols such as OPC, which we used in this work.

"FluidSim" contains a large library of basic elements that can be pneumatic, hydraulic, electrical etc.

1.3. OPC communication protocol

The OPC communication protocol is based on the connection between an OPC server and an OPC client. OPC servers are software applications (drivers) that comply with one or more OPC specifications defined in the OPC Foundation. OPC servers communicate with one or more data sources ("Data Access") on the one hand and with OPC clients on the other. In the OPC server-OPC client architecture, the OPC server is the slave, and the OPC client is the master. Communication between OPC client and OPC server is two-way, which means that OPC clients can read and write data to OPC servers [7].

The OPC Foundation defines four types of OPC servers: OPC DA, OPC HAD, OPC A&E and OPC UA server. We will use the OPC DA (OPC Data Access) server. It is an OPC Foundation specification that defines how information and data can be transferred in real time between a data source and a data sink without either knowing the underlying communication protocol of the other [8].

2. Industrial process

The aim of the work was to create an industrial process for assembling different components. Created process consisted of electrical and pneumatic components.

2.1. Initial requirements

The main objective of the industrial process is to place the covers on the bases and to sort the assembled products according to the color of the cover. An example of a product that can be assembled using such an industrial process is the Intel NUC MiniPC. Its case is made of the lower part, in which the basic computer components are located, and a plastic cover. As part of the work, we will create an industrial

IOP Conf. Series: Materials Science and Engineering 1268 (2022) 012017 doi:10.1088/1757-899X/1268/1/012017

process that places plastic covers on the already manufactured lower part of the Intel NUC computer case that we will call base.

The initial requirements of the industrial process will be shown below. Covers of three different colours (blue, green or gray) will enter the process on the first assembly line. The system will first need to check the colour of the cover that comes into the assembly process. From the assembly process, the system must then exclude gray coloured covers, as these are not suitable for assembly and represent covers with an error in a real industrial process. The blue and green covers will need to be assembled on the bases coming into the process on another conveyor belt. After assembly, the system will need to sort the assembled product by cover colour into its chute conveyor.

2.2. Industrial process drawing

Figure 1 shows a drawing of the industrial process with all actuators and sensors. Actuators are marked with blue colour and labelled AX, where X represents the number of the actuator. Sensors are marked in red and labelled SX, where X represents the sensor number.

The industrial process will consist of the following actuators: 6 pneumatic cylinders (A1, A2, A3, A4, A12 and A13), 4 conveyor belts (A5, A6, A10 and A11) and robotic XZ manipulator (A7-movement in horizontal X direction, A8-movement in vertical Z direction and A9-pneumatic suction).

For the successful operation of the industrial process, sensors needed to be added to the system. The following sensors were used: 2 photoelectric sensors for colour detection (S8 and S19), 7 photoelectric sensors for detecting objects on conveyor belts (S0, S1, S7, S11, S14, S18 and S20), 3 buttons (S15-Start button, S21-Stop button and S22-Reset button), 8 magnetic sensors for detecting the final position of the piston rod in the pneumatic cylinder (S3, S4, S5, S6, S19, S10, S12 and S13) and 3 sensors located inside XZ manipulator (S2-object detection, S16-X direction movement and S17-Z direction movement).

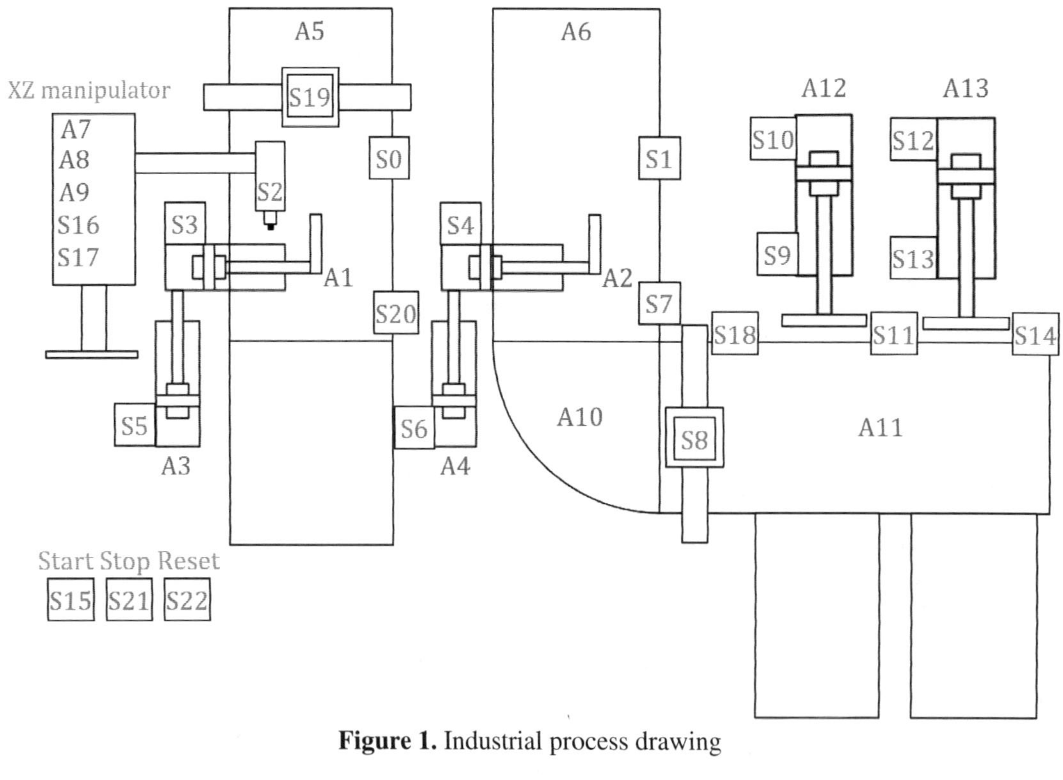

Figure 1. Industrial process drawing

IOP Conf. Series: Materials Science and Engineering 1268 (2022) 012017 doi:10.1088/1757-899X/1268/1/012017

2.3. Description of the operation of the industrial process

At the beginning, all pneumatic cylinders are in their initial positions. Cylinders A3, A4, A12 and A13 are located in the inner end position. Cylinders A1 and A2 are located in the outer end position. When the START button is pressed, the conveyor belts A5 and A6 turn on. Covers for assembly arrive on the A5 conveyor belt to the workstation. Bases arrive on the A6 conveyor belt to the workstation. When the covers and bases on both belts arrive at the assembling station, as detected by the SO and S1 sensors, the pneumatic cylinders A1 and A2 clamp them in place for assembly. Conveyor belts A5 and A6 are switched off at the same time. When the cover and base are clamped, as detected by sensors S3 and S4, the XZ manipulator moves down in the -Z direction to the lid. When sensor S2 detects the lid, the XZ manipulator pneumatically grips the lid with actuator A9. Then the XZ manipulator moves up in the Z direction and right in the X direction, above the position where the base is located. When the XZ manipulator is located over the base, it moves down in the -Z direction and releases the cover by turning off A9. The XZ manipulator and cylinder A2 then return to the initial position. The cylinder A4 then moves up in the Z direction and clears the track on the belt A6, which is detected by the sensor S6. A2 conveyor is then switched on. The S7 sensor detects whether the assembled product has left the workstation. When this happens, the A4 cylinder returns to its initial position. At the same time, the angular conveyor belt A10 turns on and transports the product to the sensor S18. When the S18 sensor detects the product, it is located below the S8 colour sensor. S8 then detects the colour of the cover on the product. Conveyor belt A11 is then switched on. Pneumatic cylinders A12 and A13 sort products according to the colour of the cover. If the cover is blue, belt A11 turns off when sensor S11 detects the product. At the same time, the cylinder A12 moves to the outer position and pushes the product from the conveyor belt A11. When A12 reaches the outer end position, which is detected by sensor S9, it returns to the initial position. If the detected product has a green cover, belt A11 is not switched off at sensor S11 but at sensor S14. When sensor S14 detects the product, A11 turns off. At the same time, the pneumatic cylinder A13 moves to the outer end position and pushes the product from the conveyor belt. When A13 reaches the outer final outer position, which is detected by sensor S13, it returns to the initial position. A sensor S19 is installed on the conveyor belt A5, which detects the color of the covers coming into the system. If a blue or green cover is detected, the execution of assembling process continues as described above. However, if gray cover is detected, the pneumatic cylinder A3 moves up and clears the path on the belt A5. The gray cover, which is not being assembled, moves past the workstation and is eliminated from the process. When the S20 sensor detects that the product has left the workstation, the A3 returns to its initial position.

Due to the large number of actuators and sensors, the entire system was divided into five smaller systems for easier transparency. The entire industrial system consists of the following smaller systems:

- System 1: Transporting and clamping the covers on conveyor belt A5.
- System 2: Transporting and clamping the bases on conveyor belt A6.
- System 3: Conveying assembled products on conveyor belt A10.
- System 4: Installing the cover on the base and transporting the assembled product down the conveyor belt A6.
- System 5: Conveying assembled products on conveyor belt A11 and sorting by color.

3. Virtual industrial process creation

The components used in the creation of the virtual industrial process in "Factory I/O" were divided into 4 groups: actuators, sensors, structural components and components for generating elements.

3.1. Actuators

The following actuators were used in the production of the industrial process: 3 straight conveyor belts, 1 angular conveyor belt, two-axis manipulator, 2 position bars and 2 large pneumatic actuators.

Straight conveyor belts are used to transport products with a small mass. They consist of supporting elements, a belt along which the products are transported and an electric motor that takes care of the movement of the belt. To create the process, 3 flat conveyor belts 4 m long were used. The electric motors on the conveyor belts are controlled digitally, the maximum rotation speed of the conveyor belt will be 0.6 m/s.

An angular conveyor belt works in the same way as a straight conveyor belt. The angular conveyor belt serves to transfer products at an angle of 90°.

The two-axis manipulator is used for gripping and transporting products in the X and Z axes. It consists of two actuators for movement in the aforementioned directions and one pneumatic actuator for gripping elements under the actuator head. The two-axis manipulator implemented in Factory I/O can operate in three different modes of operation: digital, analog, and digital-analog. In our work, we used a two-axis manipulator that works with a digital mode of operation, where the length of movement in the X direction is equal to 1.125 m, and in the Z direction 0.625 m. The maximum speed of movement of each axis was 2 m/s.

The position rod consists of two smaller pneumatic cylinders, which are positioned at an angle of 90°. The first pneumatic cylinder can move in the X direction. Its task is to clamp and position the products on the assembly line. The length of movement it can make in the X direction is 0.48 m. The second pneumatic cylinder can move in the Z direction. Its task is to release the path on the conveyor belt by lifting the first pneumatic cylinder. The length of movement that the pneumatic cylinder can make in the Z direction is 0.373 m. In our work, we used 2 position bars.

The large pneumatic cylinder consists of a pneumatic cylinder and a servo valve. Its task is to push elements from the assembly line. It can operate in analog or digital mode. In our work, we used two large pneumatic actuators with a digital operating mode, where the stroke length is 0.9 m.

3.2. Sensors

The sensors we used in the creation of the virtual industrial process are as follows: 3 buttons, 8 photoelectric sensors, 2 color sensors, 8 magnetic sensors and 2 motion sensors.

Buttons are electrical switches that serve to interact between the user and the industrial process. In our work, we used 3 buttons: the Start button to start operation, the Stop button to stop the system operation, and the Emergency Stop button.

Photoelectric sensors detect objects located in front of them. The detection range they can achieve is from 0 to 1.6 m. We used 8 photoelectric sensors in our work. 7 of them are installed next to the conveyor belts and detect objects traveling along the conveyor belts. One photoelectric sensor is placed on the gripping head of the two-axis manipulator and detects objects located under the head of the manipulator.

The color sensor is used to detect the color and type of the object located below it. The detection range is from 0.3 m to 2 m. The sensor allows 3 different modes of operation: digital, numerical and ID mode. The sensor can be assigned individual types and colors of the objects it will detect. In our work, we used 2 color sensors. The first color sensor detects the blue color of the lid on the already assembled product. The second color sensor detects gray covers arriving at the workstation prior to assembly.

Magnetic sensors act as end switches to detect the position of the piston rod in pneumatic cylinders. In our work, we used 8 magnetic end position sensors. Magnetic sensors are already implemented in the actuators described in chapter 3.1. Each position rod contains 2 magnetic sensors. The first sensor detects when the product is clamped by the position bar, and the second sensor detects whether the path on the conveyor belt is free. Each large pneumatic actuator contains 2 magnetic sensors. One sensor detects the outer end position and the other the inner end position of the piston in the pneumatic cylinder.

3.3. Structural components

Structural components have been used to assist in development of the industrial process and to facilitate the placement of sensors and actuators in appropriate locations within the system. The following structural components were used in the construction of the industrial process: 2 metal supports for color sensors, 7 corner brackets for photoelectric sensors, 3 chutes for sending products from the conveyor belts, electrical box for buttons, bracket for electrical box, 8 fences for system security, 6 small metal platforms.

3.4. Generating elements

When simulating a virtual industrial process, we used components that generate the products used in the industrial process. For this we used a special component in Factory I/O called Emitter. We needed to specify which components Emitter should generate. In addition, the minimum and maximum generation time of the components must be determined. Depending on the configuration, the Emitter will generate components accordingly if there are no components in its range. For the successful operation of the industrial process, we also needed a component that removes the products at the end of the industrial process. For this we used a component in Factory I/O called Remover. This component removes the products at the end of the industrial process when the product enters its area.

3.5. Entire virtual process

The entire virtual industrial process created in Factory I/O is shown in Figure 2.

Figure 2. Virtual industrial process created in Factory IO

3.6. Defining input and output variables

To successfully control our industrial process, we needed to define the sensors and actuators described in the previous subsections as input and output variables. These variables were then used to communicate with the controller via a specific communication protocol. The Factory I/O software tool enables connection to controllers via various communication protocols. We used the OPC Client DA/UA communication protocol to control our industrial process.

4. Development of the controller

This chapter describes the control implementation with the FluidSim software tool. We have attempted to successfully control the industrial process developed in Chapter 4 with the controller. The controller uses a sequential control chain for control.

IOP Conf. Series: Materials Science and Engineering 1268 (2022) 012017 doi:10.1088/1757-899X/1268/1/012017

4.1. Defining variables

We used the Fluidsim software tool to create the controller. We used the components located in the component library under the GRAFCET group. First, we defined the variables that we use in the control. We used 23 variables for sensors and 14 variables for actuators to control our industrial process. The GRAFCET I/O component was used to define the variables. 8 input and 8 output variables can be defined to this component. Since we needed more variables to control the industrial process, we used 3 GRAFCET I/O components.

We have to connect the values of the variables to the OPC server. Thus, we will be able to send variable values from the control in FluidSim to the virtual industrial process in Factory I/O and vice versa.

FluidSim Input Port and FluidSim Output Port components were used to connect FluidSim to the OPC server. These components are located in the FluidSim library under EasyPort/OPC/DDE. Each of these components has 8 connectors. The FluidSim Input Port component is used to read sensor values in an industrial process via the OPC communication protocol. The FluidSim Output Port component is used to send actuator values via the OPC protocol to the virtual industrial process. We used 3 FluidSim Input Port components and 3 FluidSim Output Port components for the control of industrial process. In/Out components were connected to the GRAFCET I/O components as shown in Figure 3.

4.2. Sequential control chain

The industrial process was controlled using a sequential control chain. Since the created virtual industrial process is complex, the entire process was divided it into 5 subsystems. The sequential control chain was created for each subsystem.

We used GRAFCET components Step, Transition and Action for programming of our controller. With the Step component we defined the steps of sequential control, with the Transition component we defined the conditions for transition from one step to another, and with the Action component we defined the actions that take place in a certain step.

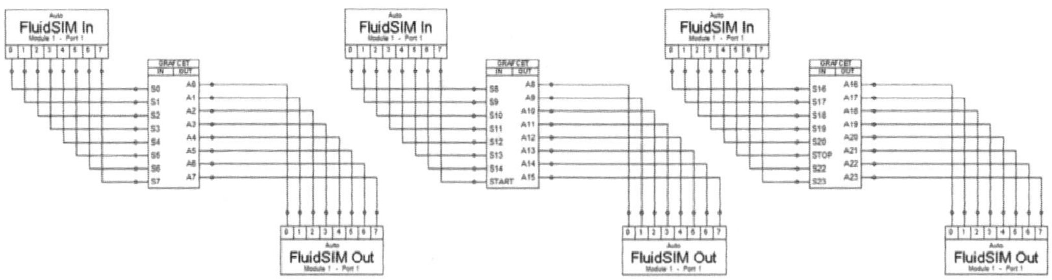

Figure 3. FluidSim connection components

5. Connection part

This chapter describes the configuration settings of the Festo Didactic EzOPC OPC server.

At the beginning, we defined the OPC server and made the connection using a test system with a smaller number of inputs and outputs. In this way, we determined whether communication between Factory I/O and FluidSim is possible. In the following, we transferred the configuration settings to a more complex industrial process, presented in the previous chapters.

5.1. Defining the OPC server

Festo Didactic EzOPC is a software tool used to define and set up an OPC server. It enables establishing a connection between different OPC clients.

In order to successfully set up the server, it was first necessary to select the appropriate process data type and the controller type to be used. We used Process simulation and CIROS/COSIMIR as the process data type and Controller and FluidSIM as the controller type for our process. The definition of the OPC server is shown in Figure 4.

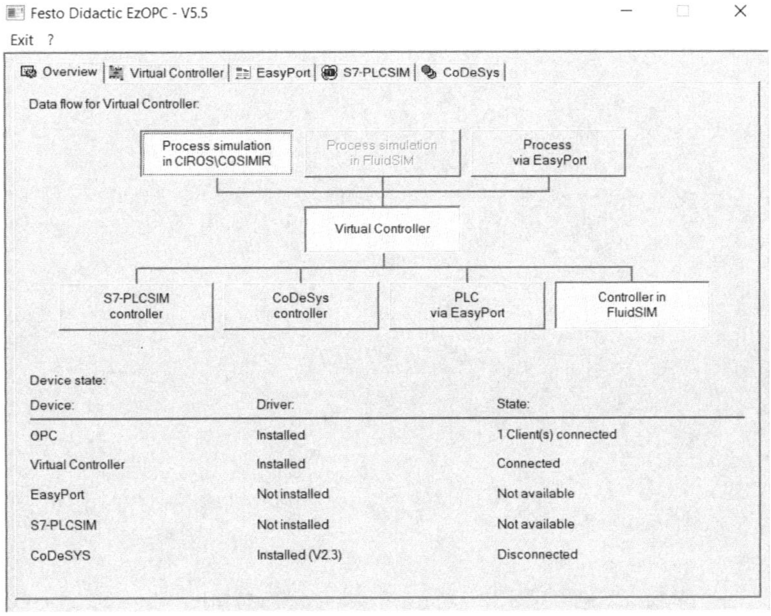

Figure 4. Definition of the OPC server in Festo Didactic EzOPC

5.2. *Connection between Factory I/O and FluidSim on test system*

We first created the test system in the Factory I/O program. In the program, we added the Start Button component to the system, which consists of a button and a light. Then, within the Factory I/O program, we defined the use of OPC communication. After setting the configuration, we determined the addresses for the input and output of the system. The input of the system is a light called Start Button 1 (Light), to which we assigned the address EB1Bit00. The output of the system is a button called Start Button 1, to which we assigned the address AB1Bit00 (Figure 5).

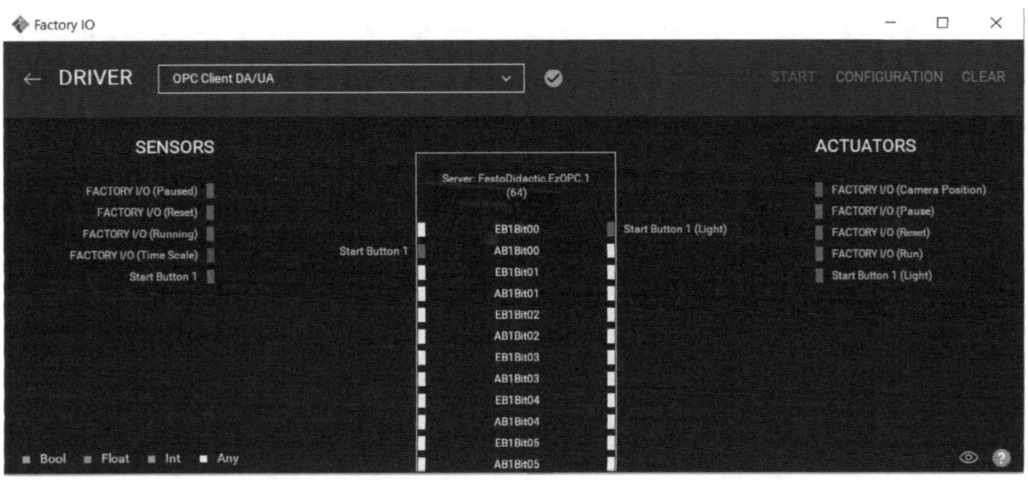

Figure 5. Definition of input and output variables

IOP Conf. Series: Materials Science and Engineering 1268 (2022) 012017 doi:10.1088/1757-899X/1268/1/012017

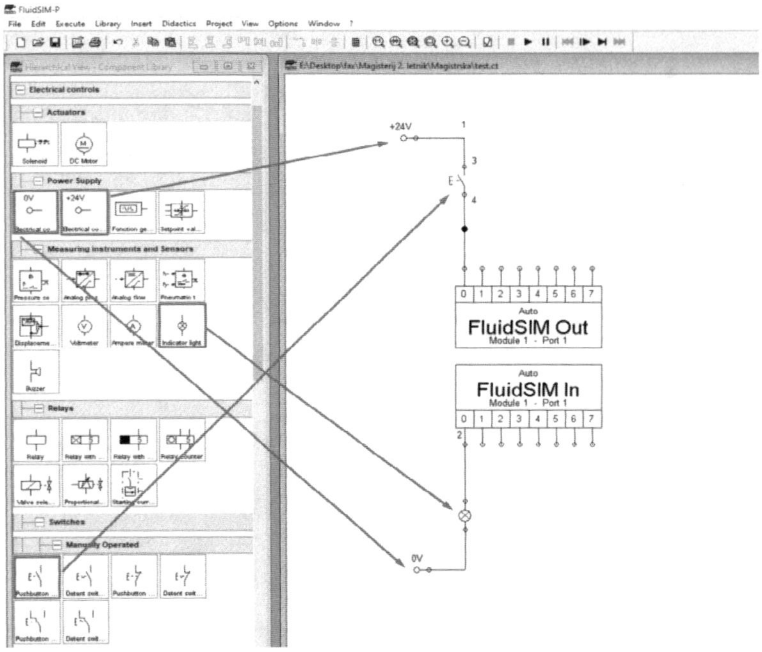

Figure 6. Test system in FluidSim

We then created a similar system in FluidSim. The FluidSim Input Port and FluidSim Output Port components, located in the EasyPORT/OPC/DDE group, have been added to the system. We also added the components Electrical connection 0 V, Electrical connection 24 V, Pushbutton (make) and Indicator Light to the system. Figure 6 shows the test system created in FluidSim.

The FluidSim Input Port and FluidSim Output Port components needed to be assigned appropriate addresses for OPC communication. For both components, we choose FestoDidactic.EzOPC.1 as the OPC server. We then defined the FluidSim Input Port address AB1. We defined the address EB1 for the FluidSim Output Port component.

After configuring the addresses, we started the simulation and checked the operation of the system. System works as follows: pressing the Start Button 1 in the Factory I/O program will turn on the light in the FluidSim program. Pressing the button in the FluidSim program will light the Start Button 1 (light) in the Factory I/O program.

Figure 7 a) shows the operation of the test system when Start Button 1 is pressed in the Factory I/O program. We can see that the light has turned on in the FluidSim program. Figure 7 b) shows the operation of the system when a button is pressed in the FluidSim program. We can see that the Start Button 1 (light) in the Factory I/O program has turned on.

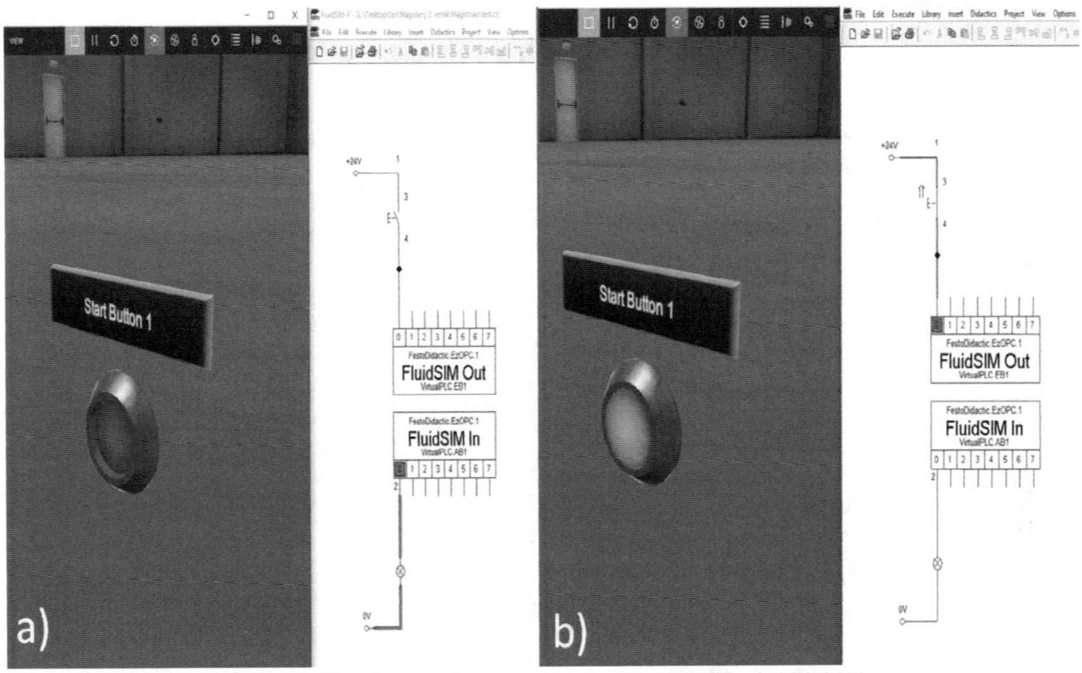

Figure 7. Simulation of test system in Factory I/O and FluidSim

5.3. Factory I/O connection with FluidSim on a virtual industrial process

After successfully simulating the test system in both software tools, we also checked the operation of the virtual industrial process created in chapter 3, which we controlled with the controller created in chapter 4. We connected the industrial process and the controller via the OPC server, just as we did for the test system described in chapters 5.1 and 5.2.

The screenshot of operation of virtual industrial process is shown on Figure 8. On the left side of the figure 8 created industrial process in Factory I/O is shown. On the right side of the Figure 8 the developed control application in FluidSim is shown. Both parts are connected in real time and are able to communicate between each other through OPC communication protocol.

6. Results

We checked the operation of the OPC connection on two systems. We first checked the operation on a test system consisting of one input and one output in Factory I/O and one input and one output in FluidSim. OPC communication was carried out via the Festo Didactic EzOPC OPC server. The connection on the simple system was successful. Switching the button in Factory I/O successfully lit the light in FluidSim and conversely switching the button in FluidSim successfully lit the light in Factory I/O.

After the successful operation of the test system, we checked the operation on a more complex industrial process, which we presented in chapter 3. We divided the entire process into 5 smaller systems, where each of the systems performed its own task. When the START button was pressed, the industrial process started to run. Initially, covers are generated on assembly line 1 and travel to the XZ manipulator. If the covers are blue or green, the positioning bar clamps them into place, but if they are gray, they leave the work area. Blue colored bases are generated on conveyor belt 2 and travel to the XZ manipulator, where the positioner clamps them in place. Once the lid and base are properly clamped, the XZ manipulator picks up the lid using pneumatic suction and transports it to the base, where it assembles the elements into an assembled product. The assembled product then travels via angled conveyor belt 3 to conveyor belt 4. 2 large pneumatic cylinders sort the assembled product according to the color of the cover. When the STOP key is pressed, the operation of the process stops. The generation of covers and bases is performed continuously as long as the simulation in Factory I/O is running and as long as there is no element inside the element generation component.

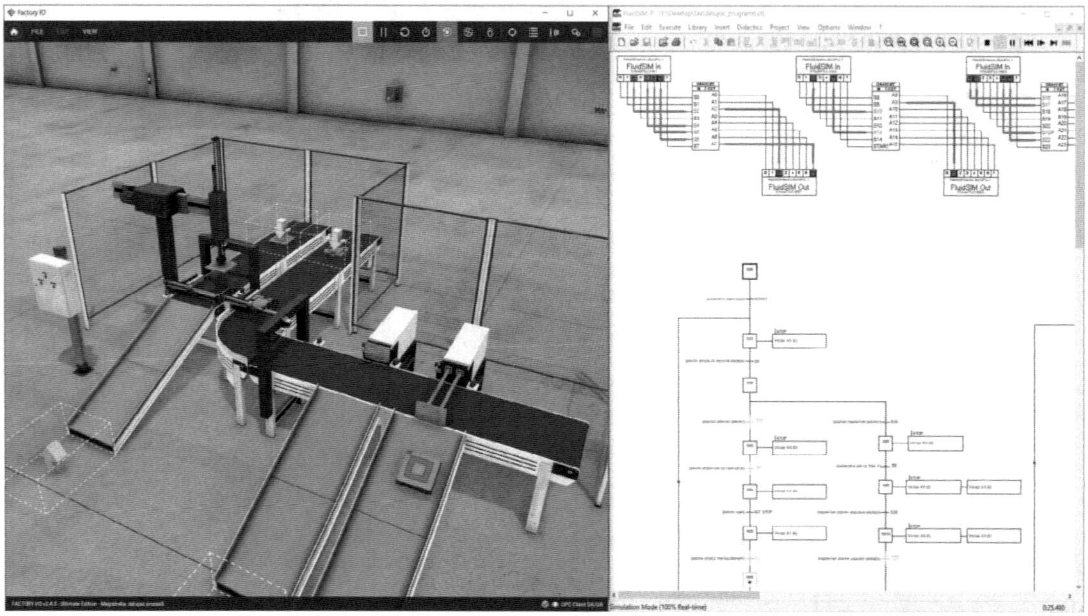

Figure 8. Simulation of virtual industrial process in Factory I/O and FluidSim

7. Conclusion

Recently, software tools for creating, controlling and simulating virtual industrial processes are increasingly used both for developing new industrial applications and for testing controls on already developed industrial applications. They allow faster developement of new industrial processes, while the planning costs are much lower, since it is not necessary to use real components. In addition, developing new systems in virtual environments is safer, as the chances of injury are much lower.

In the first part, we presented the operation of the industrial process for assembling lids and bases, sorting the assembled products and eliminating incorrect lids from the industrial process. The entire industrial process consists of 14 actuators and 22 sensors.

In the second part, we described the creation of a virtual industrial process in the Factory I/O software tool. We have described all the components used and their installation in the industrial

process. Actuators and sensors in the industrial process have been given appropriate names and addresses for input and output variables.

In the third part, we created a controller for the developed industrial process in the FluidSim software tool. The controller consists of five sequential control chains, where each control chain controls its part of the industrial process. In FluidSim, we also defined components for communication via the OPC communication protocol.

In the fourth part, we tested the operation of the OPC connection and simulated the operation of the industrial process. We first tested the operation on a test system with one input and one output. After successfully establishing OPC communication and successful simulation, we tested the operation on a virtual industrial process.

The operation of the industrial process went smoothly. The controller properly managed the industrial process with all sensors and actuators.

In our work, we achieved the goals we set for ourselves. In Factory I/O, we have accordingly created a more complex virtual industrial process. In FluidSim, we created a sequential control chain for controlling a virtual industrial process. Through the virtual controller, we connected the simulation in Factory I/O with the sequential control chain in FluidSim. We adequately tested the functioning of the

IOP Conf. Series: Materials Science and Engineering 1268 (2022) 012017 doi:10.1088/1757-899X/1268/1/012017

communication between the two software tools on the test system. Finally, we adequately tested and simulated the operation of the developed industrial process. The created virtual system could also be created in a real version, so that the created system in this part would represent a kind of digital twin of the real system.

In our work we described the creation of complex virtual industrial process using two different software tool that are connected through OPC communication protocol. We found that with the help of OPC it is possible to control virtual systems in real time and quickly find errors in the control program. The developed system made it easier to design the control program without any injuries to people or machines that could happen on real systems.

References

[1] I Ostrowski et al 2021 IOP Conf. Ser.: Mater. Sci. Eng. 1016 012005

[2] Riera, B., & Vigário, B. (2017). HOME I/O and FACTORY I/O: a virtual house and a virtual plant for control education. IFAC-PapersOnLine, 50(1), 9144-9149.

[3] Itxaso Poblacion Salvatierra, 2018, Simulation software for automation industry, Bachelor degree, University of Gävle

[4] Factory I/O User manual, 2021, Available at: https://docs.factoryio.com/manual/ [Date of access: 10.7.2022]

[5] Festo FluidSIM®4 Pneumatics User's Guide, Available at: https://www.festo-didactic.com/ov3/media/customers/1100/698528_fl_sim_p42_en_offset.pdf [Date of access: 10.7.2022]

[6] Festo Manual EzOPC 4, Available at: https://www.festo-didactic.com/ov3/media/customers/1100/04193590011200033740.pdf [Date of access: 10.7.2022]

[7] Matrikon, OPC Data Access (OPC DA) Versions & Compatibility, Available at: https://www.matrikonopc.com/opc-server/opc-data-access-versions.aspx [Date of access: 20.7.2022]

[8] Wolgang Mahnke, Stefan-Helmut Leitner, Matthias Damm, 2009, OPC Unified Architecture, Springer-Verlag Berlin Heidelberg, DOI 10.1007/978-3-540-68899-0

Experimental studies on turning process, using the Prime turning processing strategy, proposed by Sandvik Coromant

I. A. POPAN[1], N. BALC[1], A. I. POPAN[1], A. CERCIU[1]

[1] Technical University of Cluj-Napoca, Department of Manufacturing Engineering, B-dul Muncii, no. 103-105, Cluj-Napoca.

ioan.popan@tcm.utcluj.ro

Abstract. CNC turning process is a common machining process in the industrial application. The requirement on turning process is to increase the productivity and cost reductions. The paper presents preliminary research of the turning process optimization by using the Prime Turning strategy, proposed by the Sandvik Coromant company. Using a CNC turning machine, parts with complex surfaces were manufactured and two process strategies were compared. The first machining strategy is conventional roughing and finishing in the longitudinal direction and the second is the strategy developed by Sandvik company, called Prime Turning. The experimental results show that by using the Prime turning strategy and tools the productivity can be increased, with lower power consumption.

1. Introduction

Turning process is a very common machining process used in manufacturing [1]. The global competition forces the companies to manufacture at a competitive price, maintaining the products quality and a low energy consumption [2]. When turning operation is involved, the main objective is to increase the process productivity. The increasing productivity process could be translated as minimizing the machining cost and time [2], [3]. Tool wear is an important aspect when the turning process is used [3], [4]. It has a strong influence on dimensional accuracy, surface quality, tool life and productivity. On the current studies on the tool wear, the workpiece material morphology, tool wear mechanism, tool geometry and material, etc. are investigated [1], [3]–[5]. Process parameters optimization and modelling is a challenge. Machining parameters are modelled in order to improve the quality process characteristics and productivity for process monitoring. The main optimization and modelling techniques are based on empirical models, analytical model and artificial neural networks [6]. Increasing the productivity of the turning process is an intensive topic addressed in the areas of research in the field of machining. The Sandvik Coromant company proposed an innovative turning solution, called “Prime Turning”. The main advantage are: up to 50 % productivity increase and high flexibility [7]. Prime Turning concept consist on using an 25-30° entry angle in order to obtain a good chip control at a high feed rate [7], [8]. In figure 1 is presented the Prime Turning strategy as compared with conventional turning strategy.

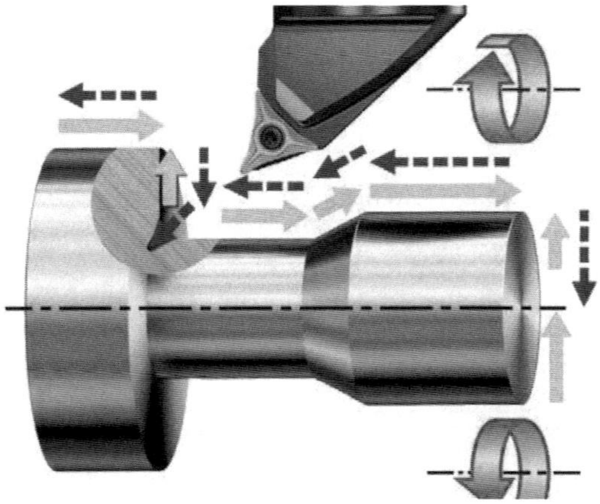

Prime turning directions
Conventional directions

Figure 1. Prime Turning strategy vs. conventional strategy [7].

The process flexibility is provided by the tool geometry. This tool with 25-30° entry angles can be successfully used for face, external and profiling turning. Roughing and finishing turning operation could be made with the same insert [7]. Stanislav Krajčoviech et. all observed in their investigations that the Prime Turning solution provides a better tool life and durability [8].

The paper presents preliminary research of the turning process optimization by using the Prime Turning strategy. The conventional roughing and finishing in the longitudinal direction strategy is compared with the Prime Turning strategy.

2. Experimental research

In this preliminary study, two strategies for processing complex surfaces by turning were compared:

- Conventional machining strategy, roughing and finishing in the longitudinal direction, programmed with the G71 cycle and
- The strategy developed by Sandvik called Prime Turning.

2.1. Design of experiments

For this comparison, a revolutionary part with complex surfaces was manufactured, a "chess part" with an external diameter of Ø30 and 40 mm length. The 3D model of the part is presented in figure 2.

Figure 2. The 3D model of the test part.

IOP Conf. Series: Materials Science and Engineering 1268 (2022) 012018 doi:10.1088/1757-899X/1268/1/012018

A number of 5 parts were machined with each strategy. The main operations of the part machining chain are roughing, finishing, and cutting off. Table 1 presents the tools and process parameters used for conventional turning process.

Table 1. The tools and process parameters used for conventional turning

Tools	Process parameters	Tool image
Roughing operation: Insert: DNMG 11 04 12-PM 4425 Holder: DDJNR 2020K 11	V_c= 317 (m/min) fn =0.35 (mm/rot) a_p= 2 (mm)	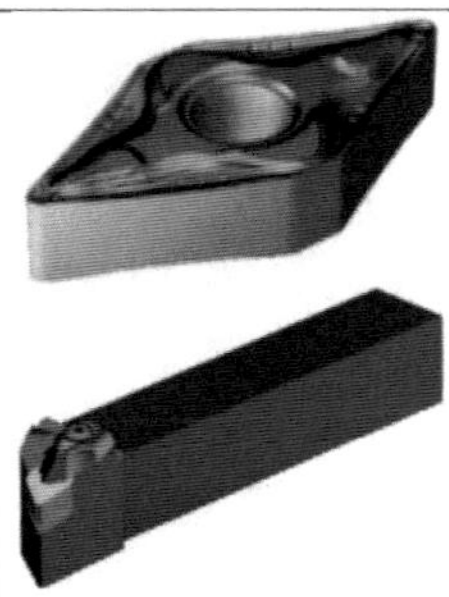 [7]
Finishing operation: Insert: DNMG 11 04 04 -PM 4425 Holder: DDJNR 2020K 11	V_c= 383 (m/min) f_n =0.18 (mm/rot) a_p= 0.5 (mm)	
Cutting off operation: Insert: QD-NE-0200-0003-CR 1135 Holder: QS-QD-RFE26C2020D	V_c= 130 (m/min) f_n =0.13 (mm/rot)	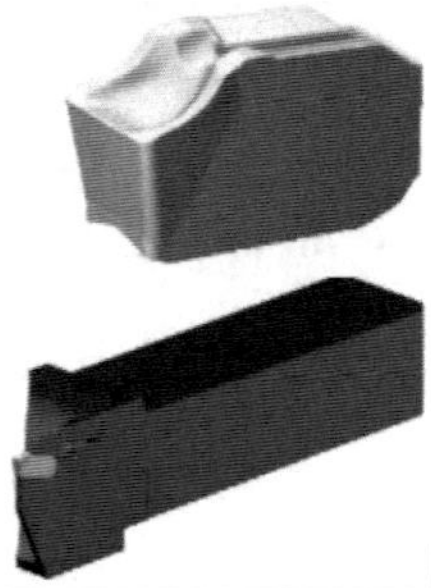 [7]

The tools and process parameters used for tests with Prime turning strategy are shown in table 2. For cutting off operation was used the same conditions with conventional turning.

Table 2. The tools and process parameters used for Prime Turning

Tools	Process parameters	Tool image
Roughing and finishing operation: Insert: DNMG 11 04 04 -PM 4425 Holder: DDJNR 2020K 11	Roughing: V_c= 445 (m/min) f_n =0.2 (mm/rot) a_p= 0.5 (mm) Finishing: V_c= 352 (m/min) f_n =0.4 (mm/rot) a_p= 0.5 (mm)	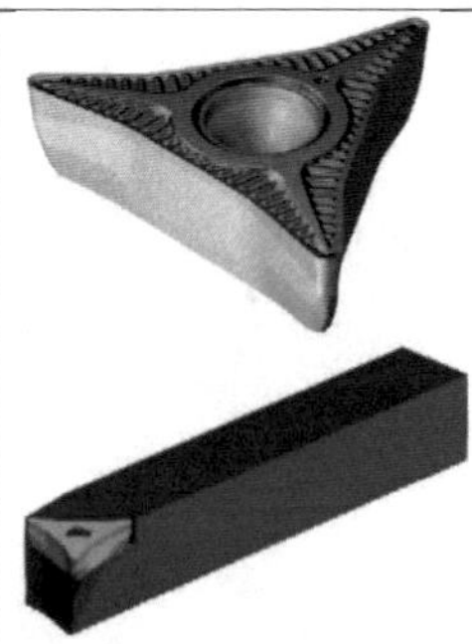 [7]
Cutting off operation: Insert: QD-NE-0200-0003-CR 1135 Holder: QS-QD-RFE26C2020D	V_c= 130 (m/min) f_n =0.13 (mm/rot)	

The process parameters used for the tests are selected based on Sandvik Coromant recommendation. Parameters were generated using Sandvik Tool Guide APP considering: the machine type and process material (C45 steel). On the experimental trails Castrol SYNTILO cutting fluid was used.

IOP Conf. Series: Materials Science and Engineering 1268 (2022) 012018 doi:10.1088/1757-899X/1268/1/012018

2.2. Equipment's and materials

The Doosan Lynx 220 CNC turning equipment was used in the experiments (fig. 3). The main technical characteristics of the Doosan Lynx 220 lathe are as follows: 15 KW spindle power, 6000 RPM maximum spindle speed, 12 positions on the turret, 320 mm travel on X- axis and 330 mm Z – axis.

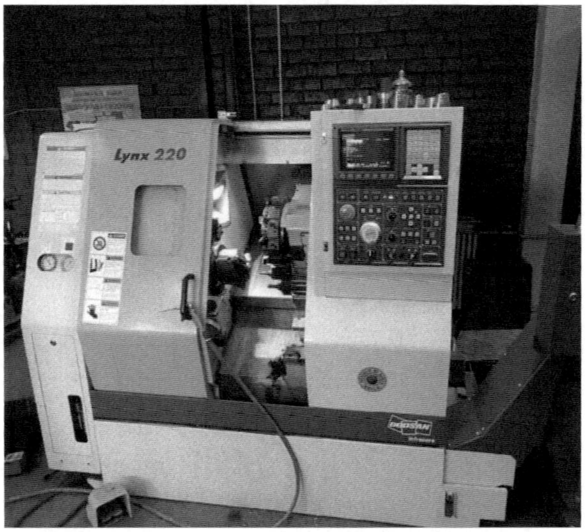

Figure 3. CNC turning machine Doosan Lynx 220.

For this experiment, steel C45 was selected, it is a common material used in industry. The 50 mm diameter bar was cut off at length of 70 mm by using Proma 250 saw (fig. 4).

Figure 4. The workpiece preparation C45 Ø50x70 mm.

2.3. CNC Programming and parts manufacturing

Writing the CNC program for the conventional strategy was made manually by using G71 cycle with longitudinal feed. The tool path is presented in the image 3.1. Prime turning strategy requires moving the tool in several directions following curvilinear trajectories that are difficult to calculate manually, requiring the use of a CAD/CAM software application. The CNC program was made using Siemens NX CAD/CAM software. The tool path for Prime turning strategy is presented in figure 5.

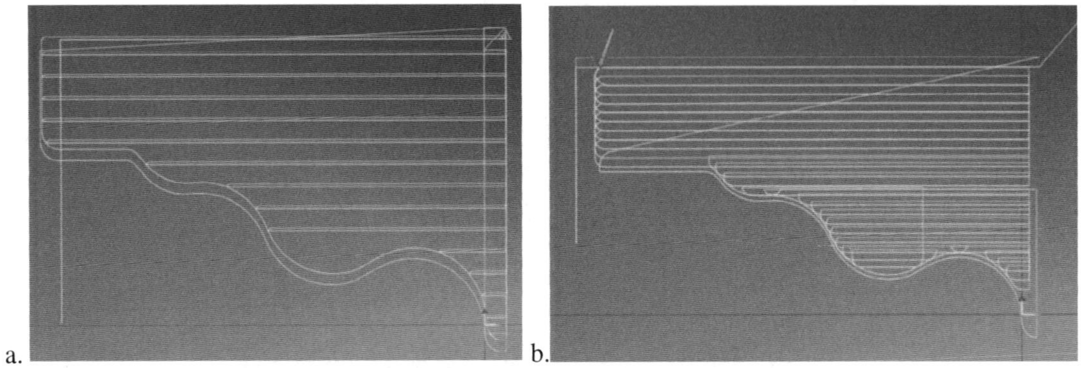

Figure 5. The tool path in case of conventional strategy (a) and Prime Turning strategy (b).

The workpiece with a diameter of 50 mm from C45 was clamped in a hydraulic clamping device, with hard jaws, in order to increase the part stiffness (fig. 6 a). In figure 6 b is showed how the tools was clamped in turret.

Figure 6 Experimental setup: a) Clamping system, b) Tool clamping system.

In figure 7 are presented images taken during the machining with Prime turning strategy. In this image is presented the feed direction during the roughing operation. In case of this cutting tool, it is moving longitudinal, in opposite direction from the clamping device.

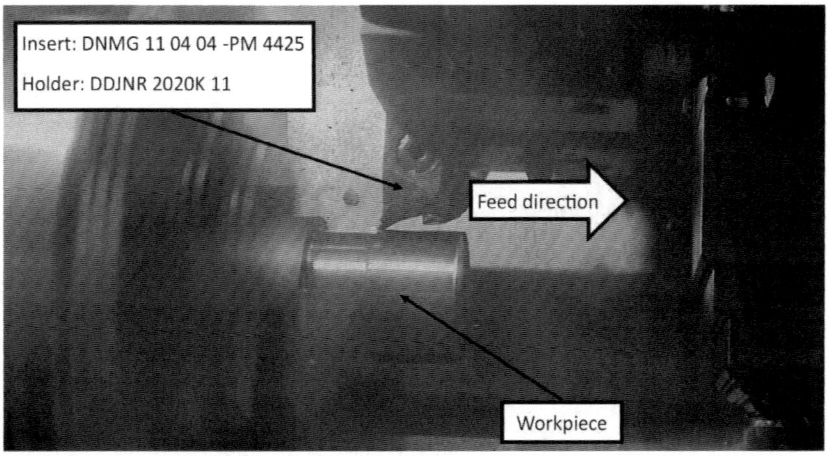

Figure 7. Images taken during the machining with Prime turning strategy.

3. Results and discussions

Figure 8 shows side by side the two pieces made with the conventional strategy vs. the Prime Turning strategy.

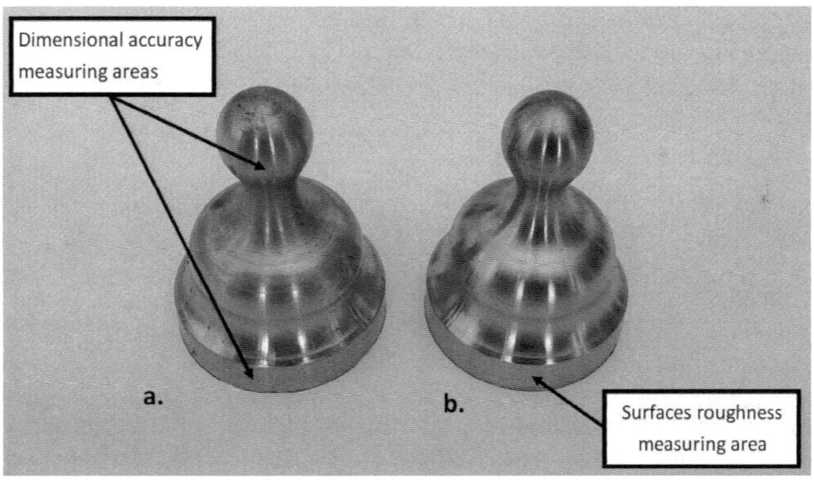

Figure 8. The parts made in experimental study: a) The part made with conventional strategy, b) The part made with the Prime turning strategy.

The parts manufactured in this experiment were dimensionally measured using the micrometre. The Ø30 mm external diameter and Ø10 mm diameter of the sphere are the measurement areas (figure 8). The dimensional measurement results are presented in table 3. The maximum dimensional deviation is +0.02. It can be concluded that, the dimensional accuracy of the process in both cases is ± 0.02 mm.

Table 3. The dimensional deviation measurement

	Conventional turning strategy	Prime turning strategy
Part 1 (mm)	+0.01	+0.01
Part 2 (mm)	-0.01	+0.02
Part 3 (mm)	+0.02	+0.02
Part 4 (mm)	+0.02	-0.01
Part 5 (mm)	+0.01	+0.02

To measure the roughness of the surfaces a roughness measuring instrument, Mitutoyo Surftest SJ-410, was used. The surface roughness was measuring on the cylindrical area of the part, as it is shown in figure 3. The results are the other quality characteristics tracked in the study are presented in table 4. It can be observed that the surface roughness is a little bit better in case of Prime turning strategy (0.9 μm) then Conventional turning strategy (1.2 μm)

Table 4. Investigation results

	Conventional turning strategy	Prime turning strategy
Machining time (min)	2.5	1.8
Surface roughness (μm)	1.2	0.9
Power consumes (KW)	8	4.9
Dimensional accuracy (mm)	±0.02	±0.02

The power consumption during the experiment was measured by the CNC Fanuc controller. In case of Prime turning strategy was measured 4.9 KW power consumption, with 3.1 KW less than the conventional strategy.

4. Conclusions

The paper presents preliminary research of the turning process optimization by using the Prime Turning strategy, proposed by the Sandvik Coromant company. Thus, by using a CNC turning machine, parts with complex surfaces were manufactured comparing two manufacturing strategies. The first machining strategy is conventional roughing and finishing in the longitudinal direction (programmed with the G71 cycle) and the second is the strategy developed by the Sandvik company called Prime Turning. Thus, the following aspects were observed:

- Using the Prime Turning strategy, a reduction in machining time can be achieved. The machining time of the part using conventional method was 2.5 minutes, and for turning the part with the Prime strategy, the machining time decreased to 1.8 minutes.
- Surface roughness was significantly better in the case of the Prime Turning strategy, from 1.2 μm it decreased to 0.9 μm.
- The power consumed in the turning process has been reduced with the Prime Turning strategy to 4.9 KW from 8 KW for conventional turning.

Using this turning strategy and Prime Turning tools can increase the productivity, and all the benefits are translated into cost reductions.
The choice of tools is therefore very important in achieving an optimal machining technology.

5. Acknowledgement

This research was supported by Horizon 2020 – DiCoMI Project, “Directional Composites through Manufacturing Innovation”, GA Nr. 778068 and by the UEFISCDI PP H2020 support project No. 77/2022.

6. References

[1] Frățilă D, Trif A 2018 *Analysis of cutting forces at dry and near-dry turning of AISI 316L stainless steel* (Academic Journal of Manufacturing Engineering) vol. 16, issue 4, pp13-19

[2] Sangwan K S and Sihag N 2019 *Multi-objective optimization for energy efficient machining with high productivity and quality for a turning process* (Procedia CIRP) vol. 80, pp 67–72

[3] Clavier F et al 2022 *Experimental analysis of the impact of an artificially generated tool wear pattern on the residual stress induced by 15-5PH steel turning* (Procedia CIRP) vol. 108, pp 394–399

[4] Tu L et al 2022 *Wear behaviours of cubic boron nitride tools with various binders in high-speed turning of compacted graphite irons* (Wear) vol. 504–505, pp 204417

[5] Liang J et al 2022 *Research on tool wear morphology and mechanism during turning nickel-based alloy GH4169 with PVD-TiAlN coated carbide tool* (Wear) vol. 508–509, pp 204468

[6] Maier M et al 2022 *Autonomous and data-efficient optimization of turning processes using expert knowledge and transfer learning* (Mater. Process. Technol.) vol. 303, pp 117540

[7] Sandvik Coromant 2022 *https://www.sandvik.coromant.com/en-gb/campaigns/primeturning/pages/default.aspx*

[8] Krajčoviech S et al 2021 *Identification of process Prime A turning when machining steel C56E2 and monitoring of cutting forces* (Transp. Res. Procedia) vol. 55, pp 605–612

AUTHOR INDEX